Radiation Survival:

How to Prepare Now and Survive Radiation Exposure

Larry and Cheryl Poole

Dedication

Dedicated to our children and grandchildren

and to those of our readers.

May they read, prepare, live confidently and survive.

Table of Contents

Disclaimer

This book is believed to contain correct information based on our personal research and preparation. We do not make any warranty, express or implied or assume any legal responsibility for the accuracy, completeness, or usefulness of any information product or process disclosed or represent that its use would not infringe on privately owned rights.

Reference to any specific commercial product process or service by its trade name or manufacturer or otherwise, does not necessarily constitute or imply its endorsement, recommendation or favoring by the authors.

The information in the book should not be a substitute for professional advice and training. The authors disclaim any responsibility for problems that may occur by following the information in this text.

Some of the information is based on new research. We encourage readers to discuss treatments with their physician and to monitor radiation neutralizer research.

Foreword

Cold War era atmospheric nuclear testing has contaminated large parts of the planet with radiation. Strontium-90 and cesium-137 were never present on the earth before the nuclear bomb testing. These isotopes have had six decades to penetrate our food supply. They will not decay for more than two centuries.

Nuclear weapon testing is not the only source of radiation exposure.

Over 90% of our nuclear plants are 20-40 years old, and are being re-licensed for another twenty years. Some are reported to exhibit apparent physical deterioration. 55,000 reports of nuclear plant licensee operational "events" have been recorded at the Nuclear Regulatory Commission.

Regional nuclear power plants are not the only concern. Accidents can spread radiation around the world.

After the nuclear plant accident in Japan on March 12, radioactive steam was released to the atmosphere. Three days later, uranium-238 was detected in air samples in Riverside, California. On March 18, it was detected in San Francisco and three days later, in Kauai, Hawaii.

Ten days after the accident, rain samples revealed radioactive cesium and iodine in Idaho. On March 23, radioactive iodine was found in rain in Albany NY, and on the following day in Tennessee and Washington state.[1] However, the EPA said the levels were so low there was no cause for concern.

---

[1] EPA, EPA Rad Net Precipitation Results, April 4 2011 report

More countries are creating nuclear weapons with a stated intent to use them to destroy their enemies. The newer, smaller nuclear weapons are portable and potentially accessible by terrorists. Reports of illicit trafficking of nuclear material continue.

In the four decades prior to nuclear testing, 1900-1940, the cancer mortality in the U.S. averaged 80 deaths per thousand. In the four decades since nuclear tests, 1950-1990, the cancer mortality in the U.S. averaged 159 deaths per thousand, double. Is it a coincidence?

Long term follow-up studies of atomic bomb survivors have confirmed an increased risk of cancer and other health effects. Solid cancers take decades to develop and chronically exposed populations continue to provide evidence of the long term low level risk.

Cancer is not the only effect of radiation. Recently, high doses have been connected to heart disease and stroke. Lower radiation doses than expected caused cataracts in the Chernobyl accident survivors. Reduced immune system functioning has been revealed in animal studies.

World governments have developed plans for a response to radiation exposure disasters. In the U.S, that planning guidance information has been shared with emergency responders. We believe ordinary Americans should know how to protect themselves and survive.

In 1962, a federal program advised citizens of radiation hazards, and community fallout shelters were stocked. Today, there is no fallout shelter program. All citizens are on their own to find shelter and to learn about radiation protection. We hope to provide you with the information you need to protect your family.

This book will provide the practical information to increase survival after all degrees of nuclear attack, nuclear accidents and radiation exposure. Regardless of the source, when a major radioactive event occurs, survivability will depend on time, (cumulative exposure to a dose), distance (significant reduction of dose with distance) and shielding (intensity is reduced by absorption and scattering by objects/walls). Shielding is the first focus, until it is safe, based on the fallout level to increase distance and reduce time exposure.

All three above factors can be maximized to your benefit, regardless of where you are at the time of the contamination. You must simply be informed about the best survival options in advance of the radiation exposure event. Also, evidence of potential radiation neutralizers have been recently reported. We share those discoveries in this book and encourage the reader to continue to monitor radiation neutralizing research.

It is our hope that this information will be valuable for preparation in case of a radiation incident, but never needed. If a radiation incident occurs, readers will have the radiation neutralizers available when use is recommended by officials or their physicians. Readers will recognize the best radiation shelters at a critical moment. Readers will know how to decontaminate themselves and the immediate environment. After reading this book, readers may be better informed than most government staffers assigned to a radiation disaster.

Chapter One

Radiation Basics

Types of Risk

A flash of light, a blast of heat and wind, and part of the city is vaporized into a mushroom cloud that contaminates the atmosphere.

We have seen the movie version of a nuclear attack. The reality is more survivable.

Survivors must know the comparative safety of immediate shelter options. The government has conducted the tests and knows the relative value of structures but this information has not been shared with the public. The government has methods to remotely monitor the radiation level after an attack.. However, those who are in shelters will not have immediate access to that information. The government will eventually have a rescue operation but in the most affected zones, it will not begin for at least 12 hours. With the information provided in this book, readers will know the best shelter option and will have a radiation emergency kit prepared. Readers will know the basics of decontaminating and potentially neutralizing the damage.

Ionizing radiation cannot be seen, felt, or smelled. Fortunately, there is a wide range of radiation detectors available for sale. There are also free plans to build an effective radiation detector.

In another city, a plume arises above a nuclear power plant and slowly descends in a cloud of steam. Area residents are transferred to public shelters with few possessions. The evacuees are dependent on authorities for safety information and for their daily living needs. Decontamination and neutralization agents may not be provided or given too late by the authorities for the maximum benefit.

In this book, we will provide suggested supplies for your own radiation emergency kit.

Until fifteen years ago, it was assumed that radiation damage was permanent and progressive. Now scientists have discovered that some common dietary supplements can neutralize radiation damage and assist in repair of cells. A sampling of radiation neutralization research is presented so that readers can investigate and discuss potential treatments with their physician.

In a rusting, abandoned manufacturing facility near a rural town, tons of radioactive materials are stored. A seasonal flood covers the ground. Later, radioactive contaminants are revealed in the water sampling.

This book will provide a description of a few documented sites of contamination. Readers can use the references to research current levels of contamination of the air and water in their region. Due to past accidents and military production, there are many unexpected areas of contamination. The reader will be able to identify specific sources of radiation and avoid unnecessary exposures.

Air, food, water and surfaces can become contaminated with radioactive elements. Those elements can be ingested or inhaled and induce radiation for the remainder of your life. It is possible to decontaminate environmental surfaces using zeolite, volcanic soil, clays and detergents. Decontamination efforts from the Chernobyl experience will be shared.

In a modest home in the suburbs, a father dies unexpectedly of lung cancer. He never smoked. No one knew that their basement family room was filled with cancer causing radiation from the natural decay of uranium underground. The second leading cause of lung cancer is from radon.

Unlike the prior three hazards, radon is a natural radiation. We cannot control its creation. However, we can remedy the contamination through effective venting, once it is recognized. 15% of our population live in homes with radon exposure.

The last chapter in the book addresses radon safety. It is a unique but deadly form of radiation. Radon requires a different type of detector and remedies.

The following section on radiation science is important to many readers to understand the potential threat. We want readers to be fully informed.

Ionizing Radiation Science

What makes ionizing radiation damaging to living cells?

Ionizing radiation is produced by unstable radioactive elements. These elements lack stability between protons and neutrons in the nucleus. The elements release excess energy in alpha and beta particles and gamma rays in order to reach stability. It is these particles and rays that emit ionizing radiation.

The particles and rays emitted are dangerous. They can inhibit cell division, break chromosomes and destroy cells and membranes of living organisms. Radioactive elements can be transferred into plants and animals through rainfall and soil contamination that is ultimately ingested or inhaled by people.

The concern about ingesting or inhaling radioactive elements internally is because the elements will continue to emit alpha, beta or gamma rays inside our bodies. These particles or rays will cause destruction until the element reaches stability.

What happens to the radioactive element when it decays?

Some elements convert directly to a non radioactive element and some to a less radioactive element before eventually becoming non radioactive. Radioactive iodine becomes non radioactive xenon. Radioactive cesium becomes non radioactive barium. Radioactive strontium decays to less radioactive yttrium. Each element will eventually change into a stable element that is non radioactive.

What is the atomic number?

The number of protons in the nucleus determines the atomic number. So far, elements with 1 to 118 protons have been discovered. Hydrogen, the lightest element, has only one proton. Helium the next lightest element has two protons. Uranium, one of the heaviest natural elements, contains 92 protons. When the number of protons changes, the element changes.

What is an isotope?

The number of neutrons in the nucleus determines the isotope. Some isotopes are radioactive and others are not, even of the same element.

Most isotopes of iodine are non radioactive. However, iodine-131, with 131 neutrons, is radioactive. Carbon is usually non radioactive, but if it has 14 neutrons, carbon-14, it is radioactive.

There are naturally occurring radioactive isotopes, including carbon-14, uranium-238 and lead-210.

Thousands of isotopes of elements have been found. 200 radioactive isotopes are produced artificially for military, energy and medical uses. The most common are; cesium-137, americium-241, cobalt-60, strontium-90 and thallium-204.

Is non-ionizing radiation dangerous?

Non-ionizing radiation is less dangerous because it does not emit cell damaging rays or particles. It can still be hazardous at high levels. Microwave and infrared radiation can heat tissue enough to cause damage. Most of the sun's energy is non-ionizing, but is still dangerous to skin tissue at high levels.

In this book assume that the term "radiation" refers to ionizing radiation.

How long do radioactive elements continue to decay?

It depends on the element. Iodine-131 will not decay to a negligible level for 81 days (10 half-lives). Strontium-90 and cesium-137 can last about 300 years or more. During that time, they will continue to emit ionizing radiation.

What is a half-life?

Half life is the time it takes for half of the radioactive material to decay. In two half-lives, one-fourth of the original radiation will remain. In three half-lives, one-eight will remain. After ten half-lives 0.1% of the original radioactive material will remain.

Amazingly, the half-life of uranium-235 is 4.5 billion years and of plutonium-244, 80 million years.

This is the reason you need to know which element and isotope have been found in air or water. This is also the reason why areas are dangerous decades after a disaster and even after residents quit worrying about the effects.

How do alpha, beta and gamma differ?

Alpha Particles

Alpha particles are relatively large and are easily deflected. They can be stopped by a sheet of paper. They pose no risk to the outside of our bodies. They cannot penetrate even dead skin cells. However, if radioactive elements that emit alpha particles are inhaled as dust, or ingested as water or in food, they can cause serious cell damage inside our bodies. The radiation will eventually decay, perhaps decades later. Most radioactive elements emitted in nuclear detonation or nuclear power plants take longer than our lifespan to completely decay.

Alpha particles were used in nuclear medicine initially, but now rarely. New cancer therapies are using alpha emitters again. Alpha emitter examples are plutonium-238, uranium, radium, radon and americium-241.

Beta Particles

Beta particles can travel across several feet in the air and through an inch of skin. They can produce radiation burns. After Chernobyl, many victims had burn injuries in their mouth, nose, eyes and skin. Beta particles can be stopped by solid material, even a sheet of plastic, but not porous surfaces. Like other types of radiation, if radioactive elements that emit beta particles are inhaled or ingested, they create a long-term health risk.

They are used in therapeutic medicine treatments. Examples of beta emitter elements are strontium-90, iodine 131, and tritium.

Gamma Rays

Gamma rays travel at the speed of light and diffuse through our bodies and most objects. They are prevalent during the initial nuclear blast. About 90% dissipates in the first seven hours after a detonation. Gamma radiation often accompanies alpha and beta particles but gamma radiation is not a particle, but a wave. It is electromagnetic. Because of the high energy, it is very harmful to our organs.

Gamma rays can travel across long distances. Gamma rays are a major cause of immediate injuries. Gamma rays can be obstructed by 90-degree shielding. Entranceways to shelter should have an L-shaped configuration and be more than 20 feet in length if possible. Each L-shaped shield will reduce gamma radiation by a factor of 10.

Gamma rays are also used in diagnostic medicine. Gamma emitter examples are cobalt-60, cesium-137 and iridium- 192. Gamma rays are also produced by the sun, although very little reaches the earth.

Are there other types of ionizing radiation?

Neutrons are only emitted during a detonation or nuclear reactor accident and quickly decay. Neutrons are unique because they can induce radioactivity in most material, including body tissue. Neutrons can easily pass through most material, causing biological damage.

Fortunately, neutrons are not as commonly encountered as other types of radiation.

The neutron bomb is a type of thermonuclear weapon. Neutrons can penetrate through thick materials, such as armor, making them useful as anti-tank weapons. This weapon is specifically designed to kill humans with radiation.

X-rays

X-rays are another form of electromagnetic radiation and very similar to gamma rays. They can be produced without a radioactive element. Isotopes that emit x-rays are iodine 125 or iodine 131. Lead or other dense material is needed to shield against x-rays.

Why is precipitation dangerous after a nuclear event?

Rain or snow will rinse out the radiation in the atmosphere and bring it to the ground. Otherwise the radiation might have blown further away. Survivors should avoid exposure to rain and snow.

What are common sources of radiation?

Naturally-occurring radiation

Naturally occurring radioactive elements include tritium, uranium, carbon-14, thorium, polonium-210, potassium-40, and radon. The carbon-14 atoms in the human body are derived from plants, which contain the same concentration of carbon-14 as the atmosphere.

Potassium-40 is the largest source of natural radioactivity in animals and humans. An adult human body contains about 160 grams of potassium-40.

The third major source of natural radioactivity is radon. Radon is a colorless, odorless, tasteless gas that occurs naturally as the decay product of uranium underground. Unlike carbon-14 and potassium-40, the risk of exposure to radiation from radon can be controlled and greatly reduced because it accumulates in enclosed areas that can be vented.

The exposure to radon gas varies widely from area to area. Radon gas from natural sources can accumulate in basements and any low, confined areas. According to the

United States Environmental Protection Agency, radon is the second most frequent cause of lung cancer, after cigarette smoking, causing 21,000 lung cancer deaths per year.[2]

A thorough review of radon will be presented in the last chapter.

Cosmic Radiation

The universe contains tremendous energy sources that create radiation that can reach earth from outer space. These energy-charged particles are referred to as "cosmic rays." The average exposure in the U. S. to cosmic radiation is about 29 mrem/year.

Cosmic rays collide with the earth's atmosphere, changing some particles. Many of these new isotopes are radioactive. Cosmic rays are responsible for the continuous production of some unstable isotopes in the earth atmosphere, including carbon-14.

We are continuously being bombarded by the shower of secondary particles created by cosmic rays. During a cross-country flight, we are likely to receive two to five mrem of radiation. A solar flare can significantly raise the radiation level. During a flare, the radiation can increase 1600 times normal levels.

The most extreme cosmic rays are called gamma-ray bursts (GRB)s. GRBs come from distant galaxies and are associated with the most intense explosions in the universe. If the earth were in direct line with the radiation

---

[2] EPA, "A Citizen's Guide to Radon". U.S. Environmental Protection Agency. January 2009.

from a GRB source within our galaxy, the radiation level could be lethal for most living beings on earth.

Fortunately, astronomers have not observed any object in our galaxy that could explode and cause a GRB event.

Household Items

Radioactive material has found its way into household items. 3% of radiation exposure is from household goods. Ionizing smoke detectors use americium-241, which is bonded to a metallic foil and sealed in an ionization chamber. As long as you don't open the sealed package, there is no radiation health risk. Americium-241 is a man-made radioactive metal with a half-life of 432 years.

There are numerous other household items that may contain radioactive elements:

- Old watches used dials painted with radium to make them glow in the dark. Bone tumors were seen in the workers who painted the dials.

- Lantern mantels – some early mantels made use of radioactive thorium.

- Jewelry – certain gems are irradiated by radioactive radium.

- Pottery – some older potteries were glazed with uranium oxide.

- Rock and mineral collections – some of these may contain natural radioactive elements.

- Lead crystal glassware may contain lead-210.

- Magnetrons containing thorium were used in microwave ovens.

- Old porcelain and bright green glass made before 1950 may contain uranium.

- Granite counter tops may contain radioactive elements found naturally in the stone.

How long has radiation been studied?

Radiological science is still a relatively new science. Ionizing radiation was first detected in 1890. Since the beginning, radiation science had the significant potential for both positive and negative outcomes.

Many researchers developed cancers from the radiation-induced DNA mutations. Thomas Edison developed an X-ray viewing machine for physicians in 1896. In 1904, one of Edison's assistants died from a carcinoma, and Edison halted the research.

Marie Curie discovered polonium and also experimented with radiation medicine. She died from the radiation exposure. Her research papers are still too dangerous to handle without protective equipment.

What industries use ionizing radiation now?

Radiation has been tested, used and defined in three separate industries; military, energy and medicine.

Most of the research and expertise come from within these industries.

Military

The production of fissile uranium and plutonium requires tremendous effort and resources. The Manhattan project, the U.S. program that produced the first atomic bomb, required 38 million board feet of lumber, 15,000 tons of silver from the U.S. treasury, and huge amounts of electricity.

More efficient methods, using ultra-high speed centrifuges, have been developed and are in use in Iran for the production of fissile material. Other newer methods, using gaseous diffusion technology, may reduce the cost of production even further. As the cost of creating material for nuclear weapons goes down, the risk of a terrorist group gaining access to nuclear weapons increases.

A Soviet physicist proposed a cobalt bomb. The cobalt would be converted into cobalt-60 by the nuclear explosion. Cobalt-60 is an extremely radioactive isotope of cobalt. This proposal was named "The Doomsday Bomb," it is unknown if it was ever built.

Since the test ban treaty of 1963, the nuclear science has largely moved from a weapon of destruction to the positive attributes in medicine and energy. The "Atoms for Peace" slogan was announced by President Eisenhower to the U.N in 1953. The International Atomic Energy Agency, (IAEA), began their mission for safe, secure and peaceful uses of nuclear sciences and technology.

Medicine

The field of medicine has promoted the benefits of radiation, especially for the treatment of cancer. Medical radiation diagnoses and therapies account for 79% of the man-made radiation exposure. In 1980, the average dose of radiation was 0.54 mSv, and by 2006, it was 3.2 mSv. It is reported that 25% of the 395 million nuclear medicine procedures are unnecessary.

The FDA issued a White Paper initiative in 2010 to reduce unnecessary radiation exposure from medical imaging. They base their concern on the doubling of the exposure to ionizing radiation in the last two decades. One study cited in the White Paper found that 29,000 future cancers could be caused by the CT scans conducted in 2007 alone.

There is no way to distinguish cancers caused by radiation from other causes. It is impossible to tell if a cancer cured by radiation was actually caused by radiation.

Energy

The energy industry has promoted the efficient production of nuclear power. Recently, they announced an expected renaissance because of the green, low-carbon emissions of nuclear power.

Why are there so many standards of radiation measurement?

Each of the radiological fields has their own language, definitions and relative benefit/risk assessment. The reports of each of these scientific fields have created a complicated language and system of measurement for evaluating radiation exposure. Four sets of radiation measurements are commonly used. In a later chapter, an explanation of each type of radiation measurement and the comparisons will be presented.

If the same type of radiation emitter is measured, the following computation can be made.

1 R = 1 rem = 1 rad = 0.01 Sv (Sievert) = 0.01 Gy (Gray)

1 Gy = 1 Sv = 100 rem = 100 R = 100 rad

What radioactive elements are used?

The following is a list of some radioactive elements, their half-life and the type of particles emitted during decay.

Element	Half Life	Particle
Cesium-137	30 years	beta, gamma
Iodine-131	8 days	beta, gamma
Strontium-90	28 years	beta
Plutonium-238	88 years	gamma, alpha
Uranium-235	millions of years	alpha, gamma
Cobalt-60	5 years	beta, gamma
Tritium	12 years	beta
Iridium-192	74 days	beta
Americium-241	458 years	alpha

| Technetium | 6 hours | gamma |
| Chromium-51 | 27 days | gamma |

Half-life is the time it takes to deplete half the radioactive material. After 10 half lives, the element is reduced by a factor of 1000.

What radioactive elements are commonly found after a nuclear accident or weapon?

Cesium 137 and 134

Cesium-137 is a radioactive isotope which is formed during nuclear fission. It has a half-life of 30 years. The isotope cesium-134 is also emitted but has a half-life of only two years.

All cesium-137 existing today is man-made. It was one of the major sources of radioactive contamination in the Chernobyl and Fukushima nuclear accidents. Cesium-137 decays by beta and gamma radiation.

Major contamination by cesium-134 and 137 occurred also from atmospheric testing of nuclear detonations in the 1950's and 1960's. Everyone was exposed to cesium because of the worldwide fallout. The current exposure from atmospheric fallout is reportedly less than one mrem and diminishes each year.

Both types of cesium can be ingested through food or drink. You can also be exposed externally by walking on contaminated soil or touching a contaminated object. Exposure to cesium is known to increase the risk of cancer. The biological interaction of cesium and potassium is a subject of animal and plant research.

Cesium eventually decomposes to barium and then is non-radioactive.

Iodine 131

This isotope of iodine-131 does not occur in nature, but is produced by man-made nuclear fission. Large amounts of radioactive iodine were released during the testing of nuclear weapons between the 1940's to 1960's. 24,000,000,000[3] curies have been released worldwide. 50,000,000 curies of iodine were released at Chernobyl.

Iodine-131 has a half-life of eight days, making it extremely radioactive, but for a relatively short time.

Iodine-131 decays into non radioactive xenon with the emission of Beta particles and gamma rays.

Iodine can be absorbed through the skin in addition to inhalation and ingestion. The body cannot distinguish the radioactive isotope from healthy iodine.

Iodine-131 can contaminate the food supply. Iodine, like other radioactive particles, settles to the ground and contaminates grass. When sheep, goats or dairy cows eat the grass, the iodine becomes concentrated in their milk.

Iodine-131 is damaging to the thyroid and has been shown to cause cancer. Chernobyl studies have revealed a high rate of thyroid cancers in children exposed to radioactive iodine.

---

[3] Agency for Toxic Substances and Disease Registry, Radiation Exposure from Iodine 131: Exposure Pathways, ATSDR, http://www.atsdr.cdc.gov/csem./csem.asp p 5

Treatment with potassium iodide within the first hour of exposure is 90% effective. Treatment is recommended in the first three hours after exposure. More information will follow in the radiation neutralizers chapter.

Plutonium 238 and 239

Plutonium is one of the most toxic substances known to humankind. External contamination is not a concern because it is an alpha emitter. Inhalation or ingestion, however, is a concern. Plutonium can move through the bloodstream after ingestion or inhalation and into bones, liver or other organs. Like strontium-90 and cesium-137, it can stay in the body for decades, exposing surrounding tissues. Tests of the urine and bones can measure the amount of plutonium in the body.

Plutonium is the heaviest element that occurs in nature. Plutonium has 94 protons. Plutonium and uranium are both involved in weapons manufacturing.

Plutonium 238 has a half-life of 88 years and emits alpha particles. It is used to power pacemakers. Plutonium-239 is the isotope most useful for nuclear weapons, since it is fissile and can sustain a chain reaction. It has a half-life of 24,200 years. Plutonium-239 has been produced in large quantity for use in nuclear weapons.

The first atomic bomb test, code named "Trinity," detonated on July 16, 1945, used plutonium as its fissile material.

The U.S. government built and operated plutonium production reactors at high-security government facilities between 1944 and 1988. There were 100 tons of plutonium produced. In addition, 100 million gallons of hazardous wastes of acids and fission products resulted from the processing.

The nuclear weapons tests in the 1950's and 1960 have left fallout of plutonium in low concentration in soils around the world. Current efforts to decontaminate the areas around nuclear weapon production facilities are in process and will be described in a later chapter.

Strontium 90

Strontium-90 has a half-life of 29 years and emits beta particles. Strontium-90 is released during nuclear accidents and detonations. It does not occur in nature. It can be inhaled or ingested through contaminated water or food.

Since the nuclear test ban treaty of 1963, strontium-90 has declined significantly. Significant strontium-90 was released during the Chernobyl disaster, but due to the large particle size, deposited nearer the site.

Strontium 90 is present in the bones in almost all living animals currently, based on autopsy results. This is amazing, considering prior to the 1940's it was not present on Earth. In 1995, the median level of strontium-90 in drinking water was 0.1 pCi per liter.[4] The maximum federal limit for Strontium 90 is 8 pCi/L which would translate to 4 mrem exposure per year.[5]

When strontium-90 is ingested, 70-80% passes through unabsorbed. The body incorporates the remainder in

---

[4] ATSDR , Agency for Toxic Substances and Disease Registry, " Toxicological Profile for Strontium," Potential for Human Exposure www.atsdr.cdc.gov/toxprofiles/tp159-c6.pdf

[5] EPA, Radiation Protection, "Strontium", ToxFAQs for Strontium, www.epa.gov/radiation/radionuclides/strontium.html

bones and teeth as if it is calcium. This increases the risk of bone cancer and leukemia. It is known as a "bone seeker."

A study found that strontium-90 was found in the baby teeth of 515 children in 1980.[6] Those children were born after the end of atmospheric nuclear tests. The investigators questioned the possible role of nuclear reactor releases.

A 2011 reported a study of 85,000 deciduous teeth collected from Americans during the bomb testing years. A sampling of 97 teeth of cancer victims revealed that those who died of cancer had significantly higher strontium-90 in their teeth.[7]

Cancer due to strontium-90 has been confirmed in animal studies.

Seemingly paradoxical, strontium-90 is used in medicine as radiation therapy implants for pituitary gland and breast and nerve tissue cancers.

Tritium

Tritium was dispersed in the nuclear weapons testing of the Cold War. It is almost always found as radioactive water. The quantity in the atmosphere peaked in 1963 and is decreasing. Nuclear power plants reportedly may release it in steam, or it may leak into the ground water. These releases are restricted by federal limits.

---

[6] Gould, JM, Sternglass, EJ, JD Sherman, "Strontium 90 in Deciduous Teeth as a Factor in Early Childhood Cancer", International Journal of Health Services, 2000, Vol 30, No. 3, 515-39 www.ncbi.nlm.nih.gov/pubmed/11109179

[7] Mangano, JJ, Sherman, JD "Elevated in vivo Strontium-90 From Nuclear Weapons Test Fallout Among Cancer Decedents: A Case-Control Study of Deciduous Teeth", Int. J. Health Serv. 2011,41(1):137-58 www.ncbi.nlm.nih.gov/pubmed/21319726

Tritium can be absorbed through the skin. It distributes in the human tissue like water. It disburses quickly and is either absorbed or excreted in urine. It has a half-life of 12 years.

Near a nuclear power plant in Russia, the water supply had three times the level of tritium than expected. There is no economically feasible treatment to remove tritium from water.

Decontamination treatment for victims is through intensive fluid replacement and urination (water diuresis).

Uranium

Uranium is the major source of energy for nuclear power plants and nuclear weapons. All uranium isotopes are unstable and weakly radioactive. It occurs naturally and is commercially extracted from uranium-bearing minerals such as uraninite. Only 1% of the original ore contains uranium, so there are significant wastes of uranium "tailings" in piles at uranium mining operations.

What elements and isotopes are usually used?

Certain isotopes are expected in certain industries.

In medicine:

Iodine 131   Technetium 99

Plutonium 238   Thallium 201

Radium 226

In research:

Carbon-14   Iodine-131

Phosphorus-32   Californium-252

Iodine-125   Tritium (H3)

In industry:

Iridium-192   Cobalt-60

Cesium-137

In weapons:

Tritium (H3)   Plutonium-239

Uranium-235   Americium-241

Uranium-238

Chapter Two

————

Health Effects
of Radiation Exposure

What happens when ionizing radiation is absorbed?

When specific types of radioactive elements are absorbed internally they may be more likely to bind to bone marrow, thyroid, liver, kidney, fat tissue or bone. The most sensitive parts of the body are stomach lining, gastrointestinal tract and white blood cells.

Strontium is attracted to bones and teeth, iodine to thyroid, radon to the lungs and uranium to the kidney. Some elements are similar to natural minerals or hormones, like calcium, iodine or potassium and are easily incorporated in the body.

Radioactivity levels are sometimes compared to dental x-rays, but this underestimates the life long potential exposure from ingestion or inhalation of the elements.

Are cells permanently damaged by radiation?

Until the mid 1990's, radiation induced damage was assumed to be caused by the death of cells, repair impossible. Since that time, it was revealed that human cells can repair the radiation damage. In some cases, the repair is complete and the cell performs normally, and sometimes it is incomplete and the cell operates improperly with DNA mutations and induces cancer decades later.

Who is investigating the cancer causing effects of ionizing radiation?

The National Research Council's BEIR Reports (Biological Effects of Ionizing Radiation), are considered the most authoritative sources of data on the health effects of radiation. Seven reports have been issued. The latest was released in 2006. A complete list of the reports appears in Appendix A.

The BEIR VII, "Health Risks from Exposure to Low Levels of Ionizing Radiation" report, updated the knowledge of radiation effects based on the continued monitoring of A-bomb survivors of Hiroshima and Nagasaki. Additional information from Chernobyl consequences were also reviewed. Based on survivor monitoring, doses of 100 to 4000 mSv were found to produce excess cancers.[8]

Is there research about other possible health effects of ionizing radiation?

The BEIR VII Report also confirmed that high doses of radiation are linked to health effects such as heart disease and stroke. [9] The link between possible inflammatory responses and low dose radiation induced cardiovascular

---

[8] National Academy of Sciences, BEIRVII: Health Risks from Exposure to Low Levels of Ionizing Radiation", p6 Executive Summary

[9] National Academy of Sciences, BEIR VII : Health Risks from Exposure to Low Levels of Ionizing Radiation", Executive Summary p 1

disease is an area of needed research, according to investigators. [10]

Animal studies have supported the potential life shortening effects of radiation. Researchers have found irradiated animals died of the same conditions as non radiated animals, but prematurely.

What have researchers found based on the Chernobyl victim monitoring?

A WHO (World Health Organization) report cited a Russian study of accident workers who had an increased risk of cardiovascular death possibly related to high radiation doses at Chernobyl. The WHO reviewers acknowledged radiotherapy patients presented the same possible link.[11]

The WHO report also noted a small increase in pre-menopausal breast cancer in the most contaminated zones Further study was encouraged.

Immune disorders appeared in children chronically exposed to radiation after the Chernobyl accident. A 2011 study tried to determine the cause of the lymphocyte depletion and T-cell and B-cell immunity disorders.[12]

The Chernobyl accident has provided evidence of thyroid cancer in children exposed to radiation. There were 5000

---

[10] Hildebrandt G, "Non-Cancer Diseases and Non-Targeted Effects", Mutat. Res., 2010, May 1, 687(1-2):73 Epub 2010 Jan 25 www.ncbi.nlm.nih.gov/pubmed/20097211

[11] World Health Organization Expert Group, "Health Effects of the Chernobyl Accident: An Overview", April 2006, www.who.int/mediacentre/factsheets/fs303/en/index.html

[12] Baleva, LS, Iakovleva, IN, Sipiagina, AE, et al, "Clinical Immunological Disorders in Children from Various Observation Cohorts Exposed to Radiation Factor During Various Stages of Oncogenesis", Radiats Biol Radioecol 2011Jan-Fed;51(1)7-19 pubmed/21520612

cases in Russia, Belarus and Ukraine in those under age 18 at the time of the accident.

The Chernobyl accident worker (liquidators)survivor monitoring has provided evidence of a doubling of leukemia risk. [13]

What is Acute Radiation Syndrome?

Acute Radiation Syndrome results from a high dose of radiation. It can be exhibited in three major syndromes, hematopoietic, gastrointestinal and cardiovascular-CNS.

If a person is exposed to more than one Gy (gray). They may exhibit a hematopoietic syndrome (blood producing). Bleeding and infections occur suddenly and lead to other effects over time because of the loss of red and white blood cells and other blood components.

If a person is exposed to 6-8 Gy, they may exhibit a gastrointestinal syndrome. They can experience severe nausea, vomiting and diarrhea. The higher the radiation dose, the faster the onset of symptoms. Fluid loss, bleeding and resulting blood infections can occur.

If a person is exposed to more than 20 Gy, they may experience Cardiovascular and CNS Syndromes. Nausea and vomiting occur in minutes. Multiple organ failure occurs, and they may lose consciousness.

---

[13] WHO, World Health Organization, Media Centre Fact Sheet, April 2006, "Health Effects of the Chernobyl Accident: An Overview" www.who.int/mediacentre/factsheets/fs303/en/index.html

*See Radiation Measurement Chapter or Appendix for measurement comparisons

What have physicians learned about radiation treatment ?

Based on reports from Chernobyl investigators, symptoms should be treated vigorously. They reported the inflammatory response of the gastrointestinal system must be a major consideration in patient treatment.

Recommended gastrointestinal treatments often include fluid replacement, antibiotics, antiviral therapy, anti-vomiting agents and anti-diarrhea drugs based on symptoms. In addition, compounds that stimulate the intestinal gut re-growth were recommended.

Bone marrow insufficiency and infections due to the immune system suppression are a major symptom of the acute radiation syndrome.

Bone marrow transplants were not successful at Chernobyl. Thirteen victims underwent bone marrow transplants but only two survived.

Potential antibiotic radio-protectors of the hematopoietic system in a radiation emergency are tetracyclines and fluoroquinolones.[14]

Are there general medical guidelines for treatment of acute radiation syndrome?

The recommended treatment of acute radiation illness often includes: analysis of urine, complete blood counts, electrolyte monitoring, repeated radiation measurements, fecal samples or whole-body radiation counts, and

---

[14] Kim, K, Pollard, JM, Norris, AJ, et al, "High Throughput Screening Identifies Two Classes of Antibiotics as Radioprotectors: Tetracyclines and Fluoroquinolones", Clin Cancer Res, 2009 Dec 1:15 (23) 7238-45 pubmed/19920105

consultation with radiation experts. Radiation experts may recommend early administration of radionuclide-specific decorporation agents such as, Prussian blue or DTPA. Antibiotics and immunomodulating drugs may be used. Further details on radiation neutralizers will be provided later in this book.

Is radiation responsible for the increased cancer mortality?

The estimated lifetime risk of fatal cancer is 5% per Sv. (0.05% per rem)[15] The lowest dose for which evidence exists of increased cancer risk to humans is about 10-50 mSv for acute exposure (within 24 hours) and 50-100 mSv for protracted exposure. [16] There is little question that above 100 mSv there are serious effects on humans, including cancer.

Based on the lifetime risk estimate, if a 100 million people were exposed to 10 rem (0.1 Sv) over a lifetime, 500,000 excess cancer deaths would be expected.

At low levels of radiation, the dose can induce mutation, chromosome aberration or genetic changes in a single susceptible cell that can lead to a neoplasm. [17]

Another government report found a 5 rad -100 rad exposure increases the lifetime risk of cancer from less than 0.5% to up to 8%, respectively. [18]

---

[15]REAC/TS, Radiation Emergency Assistance Center, "The Medical Aspects of Radiation Incidents", Oak Ridge Tn, 4/19/2011, www.orise.orau.gov/reacts/medical-aspects p 45

[16] Brenner, David J, Richard Doll, et al "Cancer Risks Attributable to Low Doses of Ionizing Radiation: Assessing What We Really Know, www.pnas.org/content/100/24/13761.full August 29, 2003 p1, pubmed 14610281

[17] Upton, AC, "Environmental Standards for Ionizing Radiation: Theoretical Basis for Dose-Response Curves", Environ. Health Perspect; 1983 Oct;52;31-9 pubmed 1569362

In documented survivor monitoring, thyroid cancers were noticed in about two to four years. Leukemias were noticed in five years, and solid cancers at least a decade, usually two decades, after significant radiation exposure.

Moderate to high doses of radiation increases the risk of cancer in most organs. Thyroid, breast, lung and leukemia risk estimates are fairly accurate and found at relatively low doses. [19] (Less than .2 Gy)

Cancers associated with high-dose radiation exposure (greater than 50,000 mrem) include leukemia, breast, bladder, colon, liver, lung, esophagus, ovarian, multiple myeloma and stomach.

The cancer rate mortality has doubled since atomic research began, when comparing the four decades before and after nuclear bomb testing.

Has anyone monitored populations exposed to nuclear weapon testing?

More recent investigations have followed affected populations living downwind from nuclear weapon testing sites or nuclear accident sites. The radiation exposure levels have been lower in these groups than previous A-bomb survivor investigations. The health outcomes have not been consistent. The following are results of cancer

---

[18] National Security Staff Interagency Policy Coordination, Subcommittee fro Preparedness and Response to Radiological and Nuclear Threats, "Planning Guidance for Response to a Nuclear Detonation 2nd Edition, www.remm.nlm.gov/PlanningGuianceNuclear Detonation.pdf, p 82

*see Radiation Measurement Chapter

[19] Ron, E, "Ionizing Radiation and Cancer Risk, Evidence from Epidemiology", Radiat. Res. 1998 Nov; 150(5 Suppl): S30-41 pubmed/9806607

studies of populations exposed to nuclear weapons testing:

- A study of leukemia in children who lived in Utah but were exposed to Nevada atomic bomb tests was conducted. The greatest risks were those in a high fallout dose group that were less than 20 years old at the time and who had died before 1964. [20]

- Another study of schoolchildren, exposed to fallout from Nevada's test sites from 1951 to 1958, found a statistically significant excess of thyroid neoplasms. [21] Investigators re-evaluated the results a decade later and found an increased risk of thyroid neoplasms and autoimmune thyroiditis up to 30 years after exposure.[22]

Has anyone monitored the populations living near nuclear power plants?

Studies of area residents and proximity to nuclear power plants and cancer have been conducted with mixed conclusions;

- There were statistically significant dose response trends in radioactive emissions and leukemia cases

[20]Stevens, W., Thomas, DC, Lyon, JL, Till, JB, Kerber, RA, et al "Leukemia in Utah and Radioactive Fallout From the Nevada Test Site: A Case Control Study", J. Am. Med. Assoc. 1990 Aug 1 264:585-591 pubmed/2366297

[21]Kerber, RA, Till JE, Simon, SL, Lyon, JL, et al, "A Cohort Study of Thyroid Disease in Relation to Fallout from Nuclear Weapons Testing", JAMA, 1993, Nov3; 270(17)2076-82pubmed/8411574

[22] Lyon,JL, Alder, SC, Stone, MB, Scholl, A, et al, "Thyroid Disease Associated with Exposure to the Nevada Nuclear Weapons Test Site Radiation: A Reevaluation Based on Corrected Dosimetry and Examination Data, Epidemiology, 2006 Nov;17 (6) 604-14, pubmed 17028502

in adults living in 22 towns near a Massachusetts nuclear plant from 1978 to 1986.[23]

- However, a National Cancer Institute study in 1990 looked at 107 U.S. counties and found no excess risk from any cancers near nuclear facilities. [24]

- Breast cancer mortality rates from 1984 to 1988 were correlated with airborne releases from nuclear power plants between 1970 and 1987. The study included nine census regions where there were power plants. [25] The investigators suggested short term fission products in drinking water and fresh milk as a possible cause.

- Thyroid cancer incidence did not increase in the county in which Three Mile Island nuclear power plant is located. However, an adjacent county showed a trend toward increasing thyroid cancer beginning in 1995. Another contiguous county showed a significant increase in thyroid cancer beginning in 1990. The investigator states it does not prove a causal link. [26]

- A study of 32,135 people exposed to low level radioactivity near Three Mile Island from 1979 to 1992 found total mortality and heart disease were

[23] Morris, MS, Knorr, RS, "Adult Leukemia and Proximity-Based Surrogates for Exposure to Pilgrim Plant's Nuclear Emissions", Arch Environ. Health, 1996 Jul-Aug;51(4):266-74, pubmed 8757406

[24] Jablon, S, Hrubec, Z, Boice, JD, "Cancer in Populations Living Near Nuclear Facilities, A Survey of Mortality Nationwide and Incidence in Two States"JAMA1991 Mar20;265(11)1403-8pubmed/1999880. 265:1403-1408 (1991)

[25] Sternglass, EJ, Gould, JM, "Breast Cancer:: Evidence for a Relation to Fission Products in the Diet", Int. J. Health Serv.1993:23: (4)783-804 pubmed/ 8276535

[26] Levin, RJ, "Incidence of Thyroid Cancer in Residents Surrounding the Three Mile Island Nuclear Facility", Laryngoscope, 2008, Apr;118 (4):618-28, pubmed 18300710

significantly elevated. However, the investigators found when "confounding variables" and natural background radiation were controlled, the elevations in heart disease were not apparent. There was also a significant linear trend for breast cancer risk and gamma exposure. The researchers reported it was unlikely due to radiation exposure on the day of the Three Mile Island accident. [27]

- A second study based on the above population was published in 2003. The report stated there were some dose-response relationships with specific cancer incidences that cannot be definitively excluded.[28]

Researching cause and effect over a long time period and large population groups is always difficult. A recent study found that county boundaries are unsuitable for a cancer mortality investigation because of the varying geographic boundary lines, and insufficient populations.[29]

What is the Petkau Effect?

Dr. Abram Petkau, a nuclear researcher in Manitoba Canada found long term, low level radiation was as

---

[27] Talbott, EO, Youk, AO, McHugh, KP, et al, "Mortality Among the Residents of the Three Mile Island Accident Area; 1979-1992, Environ Health Perspct. 2000 Jun;108 (6);545-52 pubmed 10856029

[28] Talbott, EO, Youk AO, et al, "Long-term Follow-up of the Residents of the Three Mile Island Accident Area: 1979-1998", Environ Health Perspect., 2003 Mar;111(3):341-8 pubmed 12611664

[29] Cochran, Thomas B, "Limitation of Cancer Ecologic Studies of Populations near US Nuclear Plant Sites", Natural Resources Defense Council, Inc. March 10 2011, Bethesda MD, www.nrc.gov/public-involve/conference-symposia/ric/past/2011/docs/abstracts/cochranreport

destructive to cell membranes as short term high levels of radiation. He reported his findings in 1972 that 1 millirad/min radiation for nearly twelve hours had the same effect as 26 rads/min had in about two hours. Researchers are still debating the causes but believe it is related to the overwhelming and damaging free radical production of radiation. The potential implication of low-level radiation impact on our health is tremendous. This has led to research in antioxidant radiation treatments. These will be discussed later in the chapter on Radiation Neutralizers.

What is the Bystander Effect?

The newest research on low dose radiation concerns the "bystander effect." This theory proposes that cells do not necessarily have to be hit directly by radiation to be affected.

Is there any proof of health effects of low level radiation?

The mechanism of disease caused by low-level radiation is still not very well understood. Ionizing radiation has sufficient energy to change the structure of molecules, including DNA within the cells of the body.

At doses of less than 100mSv it is not statistically possible to evaluate cancer risk. However, at a dose of 100mSv, (.1Sv), 1% of those exposed would be expected to develop a solid cancer or leukemia. [30]

---

[30] National Academies of Sciences, BEIR VII Health Risks from Exposure to Low Levels of Ionizing Radiation", Executive Summary p 8

In the BEIR VII report, "Health Risks from Exposure to Low Levels of Ionizing Radiation",

- the definition of low level radiation was near zero to about 100mSv (0.1 Sv). Americans are exposed to annual background radiation levels of about 3 mSv. The report confirmed there is a linear dose response between exposure to ionizing radiation and solid cancers in humans.

- Risks for most solid tumors were 50% greater for women than men.[31]

- A 10-15% greater cancer risk than previously reported was considered in some populations. [32]

- There were increased risks for cardiovascular, respiratory and digestive disease effects of radiation.

Is there a health risk to nuclear plant workers?

Nuclear workers have provided an affected population for evaluation of low level radiation risk. Those retrospective studies have sometimes shown evidence of disease connections but are often dismissed because of confounding factors and variations in radiation dose. Confounding factors are susceptibilities in the population to illness and other non-radiation factors.

---

[31]Institute for Energy and Environmental Research, "Cancer Risks for Women and Children Due to Radiation Exposure Far Higher Than Men, press release July 7 2005, www.ieer.org/comments/beir/beir7pressrel.html

[32] Institute for Energy and Environmental Research, "Cancer Risks for Women and Children Due to Radiation Exposure Far Higher Than Men, press release July 7 2005, www.ieer.org/comments/beir/beir7pressrel.html

- A study of Hanford, WA, nuclear plant workers found a greater sensitivity to radiation induced cancer after age 50. The risk doubling dose was 26 rem at age 58, but only 5 rem at age 62 and over age 62, only 1 rem doubled the cancer risk.[33]

- An IARC (International Agency for Research on Cancer) review of almost 96,000 nuclear industry workers in three countries, US, UK, and Canada, found excess risk for death from leukemia, but not all cancers. The report summarized the range of cancer risk approximates the existing estimates of risk.[34]

- A case-control study of workers at four nuclear facilities found multiple myeloma was associated with low level ionizing radiation at older ages. The dose response increased with adult ages over 45.[35]

Are there occupational exposure limits?

Radiation occupations also include medical radiology, aerospace pilots and attendants, and uranium mining. The limit of exposure for these populations is 5 rem (50mSv) per year. In the European Union, the limit for pregnant flight attendants was set at 1 mSv in 1996.

---

[33] Kneale, GW, Stewart AM. "Reanalysis of Hanford Data: 1944-1986 deaths", Am J. Ind. Med., 1993 Mar;23(3);371-89 pubmed/8503458

[34] Cardis, E, Gilbert ES, Carpenter L, Howe G, et al, "Effects of Low Doses an Low Dose Rates of External Ionizing Radiation: Cancer Mortality Among Nuclear Industry Workers in Three Countries" Radiat. Res. 1995 May : 142 (2):117-32 pubmed 7724726

[35] Wing, S, Richardson, D, Wolf, S, Mihlan G, Crawford-Brown, D, Wood, J, "A Case Control Study of Multiple Myeloma at Four Nuclear Facilities". Ann Epidemiol. 2000 Apr;10 (3):144-53 pubmed/10813507

The EPA states there is no firm basis for setting a safe level of exposure above background levels. The EPA states 5-10 rem accumulative dose leads to changes in blood chemistry.

The EPA Protection Action Guide for nuclear incidents states workers can receive up to 10 rem (.1Sv) to protect property and up to 25 rem (.25 Sv) to save a life. Greater than 25 rem can be on a voluntary basis once in a lifetime.

The US NRC (Nuclear Regulatory Commission) Annual Regulatory Limits for occupational workers;

Whole body	5 rem (50mSv)
Any organ	50 rem (500 mSv)
Lens of eye	15 rem (150 mSv)
Skin	50 rem (500 mSv)
Extremities	50 rem (500 mSv)
Fetal dose (declared pregnancy)	.5 rem (5 mSv)

The ICRP (International Commission on Radiological Protection), Publication 103 issued general recommendations in 2007 for occupational limits. They are stricter than NRC limits in these categories:

Whole body 2 rem (20 mSv)

Fetal dose .1 rem (1mSv)

The NCRP (National Council on Radiation Protection and Measurements) has set a limit to a pregnant radiation worker at .5 mSv per month.

Are there limits for the general public exposure?

The limits are designed for exposures beyond background radiation or medical application. The NRC and the NCRP recommends .1 rem (1 mSv) limit. [36]

A maximum annual limit of 5 mSv is allowed for infrequent annual exposures. However, the "infrequent annual exposures" limit presumes that the average over five years would still not exceed 1mSv.

The ICRP also has a 1 mSv limit for non occupational exposure, but the limit of eye exposure is only 15 mSv.

A negligible radiation dose is only .01 mSv.

[36]NCRP, "Recent Applications of the NCRP Public Dose Limit Recommendations for Ionizing Radiation", NCRP Statement #10, Dec 2004
www.ncrp.online.org/publications/statements

Chapter Three

Radiation Exposure

What is the average dose of radiation for ordinary Americans?

Approximate Doses of Ionizing Radiation to Individuals per year is summarized below; [37]

Annual dose	360 mrem (3.6 mSv)*
Natural sources	300 mrem (3.0 mSv)
Man Made sources	60 mrem (.6 mSv)
Full set dental x-rays	40 mrem (.4 mSv)
Watching TV	2-3 mrem (.01 mSv)

* See chapter on radiation measurement for an explanation of the various units used to measure radiation.

Are some people more susceptible to radiation exposure?

Children are ten times more susceptible to radiation-induced cancers than adults. Some people have a defective

---

[37] ATSDR, Agency for Toxic Substances and Disease Registry, "Toxicological Profile for Ionizing Radiation", Public Health Statement Ionizing Radiation, September 1999 p 12 www.atsdr.cdc.gov/phs/phs.asp

*See Radiation Measurement Chapter

gene that prevents damaged cell radiation repair. The elderly and immune compromised are also at increased risk.

What about the exposure during the Cold War nuclear testing?

The greatest source of ionizing radiation was the atmospheric nuclear detonations of weapons. Between 1944 and 1972 over 700,000 curies of radioactive iodine was released at the Hanford Nuclear Reservation in Washington State. Oak Ridge National Laboratory accounted for up to 42,000 curies between 1944 and 1956. The Nevada test site released 150,000,000 curies of iodine during tests between 1952 and 1970.[38] Large quantities of Strontium 90 and Cesium 134 and 137 were also released.

During the Cold War, national security prevented disclosure of the risks and health hazards near weapon production facilities.

What about the exposure from medical x-rays?

Exposure to x-ray radiation makes up 18% of total exposure, and nuclear medicine procedures, another 4%. The following are expected radiation exposures for medical procedures [39] [40].

Mammography   100 mrem

---

[38] ATSDR, Agency for Toxic Substances and Disease Registry, "The Legacy of I-131 in the Environment", chart "Worldwide Major Releases of Significant I-131 Occurred at the Following Locations" www.atsdr.cdc.gov/csem/csem/asp

[39] ATSDR, "Ionizing Radiation Sources of Population Exposure to Ionizing Radiation", www.atsdr.cdc.gov/toxprofiles/tp149-c6.pdf.

[40] American Nuclear Society, "Radiation Dose Chart", www.new.ans.org/pi/resources/dosechart

Upper GI tract evaluation	720 mren
CT scan chest	700 mrem
CT scan cardiac	2000 mrem
Bone scan	630 mrem
Cardiovascular screening	1400 mrem

In the U.S., a typical nuclear medicine examination gives a dose equivalent of 5 mSv. The average radiation dose per treatment ranges from .3 to 2.2 mSv. The average dose per treatment increases with age. Adults over age 65 were exposed to three times the radiation as teens per treatment. The medical radiation exposure alone has increased six-fold since 1980.

Most of the U.S. Supplies of medical isotopes are produced in Ontario, Canada. Isotopes used for imaging include; iodine-123, technetium-99, thallium-201, and xenon-133.

What is the exposure of those living around nuclear power plants?

All nuclear power plant operators are required to monitor the airborne and liquid discharges and to file a report of these discharges annually with the Nuclear Regulatory Commission (NRC). These reports are available at the NRC Agencywide Documents Access and Management System (ADAMS). Reportedly, any nuclear facility may release very small amounts of radioactivity during normal operations.

The radiation in the air, vegetation, milk and water near the nuclear power plants are recorded. Radiation doses of strontium-90 are expected to be less than 1 mrem within 30 miles of a nuclear plant.

Are there radioactive tritium discharges?

The NRC states that nuclear power plants routinely and safely release dilute concentrations of tritiated (radioactive tritium) water, but these releases are authorized and closely monitored. The releases are available to the public at www.reirs.com/effluent. In a 2006 report, the NRC stated several nuclear power plants had reported abnormal releases of liquid tritium, which resulted in groundwater contamination. [41]

The NRC states that in spite of 38 nuclear power plants reporting leaks or spills of tritium in excess of 20,000 pCi/L, no drinking water supply exceeded the allowable limit for tritium according to the EPA Safe Drinking Water standards. [42]

Are nuclear power plants safe?

Nuclear power plants are efficient and effective at producing electricity. However, personnel errors, mechanical malfunctions, weather and the combination of these factors are documented in 55,000 reportable events at U.S. plants at the US NRC website.

In the U.S., the nuclear power plants are aging; 52 of the reactors have been in operation over 30 years. Forty-two are over 20 years old. The radionuclide releases are measured by the EPA Rad-NESHAPS (National

---

[41]US NRC, Fact Sheet "Tritium, Radiation Protection Limits, and Drinking Water Standards", July 2006 p2

[42]US NRC, "List of Historical Leaks and Spills at U.S. Commercial Nuclear Power Plants", Rev 7, 14 Jun 11, http://pbadupws.nrc.gov/docs/ML1012/ML101270439.pdf

Emission Standards for Hazardous Air Pollutants), Radionuclide program through the Central Data Exchange. Access to the release data is strictly controlled for authorized users.

Nuclear power plants were designed with three levels of defense against leakage;

1) the fuel rods trap 99% of the radiation in normal operation,

2) the reactor vessel is a middle barrier with walls up to 10 inches thick to contain the fuel rods,

3) the containment is the outside perimeter that provides a third level of protection with six feet thick concrete walls.

Three Mile Island was contained at the third level. Chernobyl was not.

Nearly three million Americans live within ten miles of a nuclear power plant. The mandated emergency plans created by the utilities and the federal agencies define two emergency zones, one within 10 miles, and another within 50 miles of the nuclear plants. The emergency plans provide evacuation information to the communities in case of an accident.

There are four categories of power plant emergency warnings according to the U.S. Nuclear Regulatory Commission; Unusual Event, Alert, Site Area Emergency, and General Emergency

Do we, the writers, support nuclear power?

Yes. In order to support our economy and a healthy lifestyle, inexpensive energy is essential. The efforts of those who fight against nuclear power plant construction have increased risks. Older plants in disrepair or in poor locations continue to be used because of the political bias against new nuclear plants.

The problems already encountered can be easily remedied. Automated systems to prevent operator errors, gravity flow of coolant, and added layers of protection between the fuel and the environment would have prevented the past accidents. Higher qualifications for a reactor operator license would be reasonable. Currently anyone with a high school diploma or GED with five years of experience and one year training is qualified for a license.

What were the most dangerous nuclear power plant accidents?

The International Nuclear and Radiological Event Scale is intended to measure the relative significance of radiological incidents and accidents. The scale from 1 to 7 can be applied to any nuclear activity. Levels 1-3 are "incidents" and levels 4-7 are "accidents." The scale increases in severity by ten fold at each level. Two "Level 7" accidents have occurred at nuclear power plants, Chernobyl and Fukushima Daiichi.

Chernobyl

In April 1986, the Chernobyl Nuclear Power Plant in Russia experienced a sudden power surge during a test that caused a reactor vessel to rupture. The test was unauthorized. The operators had disabled a safety system during the test. After the problem occurred, the measurements were misread and not appropriately addressed. There were multiple operator errors. An intense graphite fire burned for 10 days. There was no fire extinguishing equipment in the world that could address the heat of the fire.

Radioactive iodine and cesium traveled in an airborne plume for thousands of miles. Strontium was also expelled locally. Winds changed, and new areas were affected. Ukraine, Belarus, Russia, Poland, Germany, Austria and Hungary were primarily affected. Finland, Sweden and Norway were also contaminated, and radioactive material was eventually detected in the U.S. and Japan.

An area 10 kms around Chernobyl is still indefinitely uninhabitable. An area 30 kms around Chernobyl will be "controlled entry" indefinitely.

The total area of radioactive cesium-137 contaminated land is 56,000 kms [43] Residents in four settlements were exposed to additional internal cesium-137 radiation from consuming milk and dairy products and wild mushrooms and berries[44]. As of April 2006, 270,000 people still lived

---

[43] Shoigu, SK, National Statement, "IEAE Updating Report on the International Chernobyl Project " Minister for Civil Defense, Emergencies and Elimination of Consequences of Natural Disasters. P53, Proceedings of

One Decade After Chernobyl,
www.iaea.org/publications/documents/infcircs/1996/inf510.shtml

[44] Shoigu, SK, National Statement, "IEAE Updating Report on the International Chernobyl Project", Minister for Civil Defense, Emergencies and Elimination of Consequences of Natural Disasters.
P55www.iaea.org/publications/documents/infcircs/1996/inf510.shtml

in strictly controlled zones with cesium readings over 555 kBq/m². [45]

According to a UN Committee, 134 liquidators (accident workers) received doses that caused acute radiation sickness and 28 died. The authorities acknowledged that more than 28 died, but those deaths could not necessarily be attributed to radiation exposure. Chernobyl workers received high radiation doses of 70,000 to over a 1,000,000 mrem. It is generally believed that 500 rem of radiation will likely be fatal. [46]

The Prime Minister of Ukraine reported in 1996 that cases of thyroid cancer had increased several times normal levels in children and adolescents. He reported that 360,000 liquidators living in Ukraine suffered from medical effects, 35,000 were invalids.

In Russia, thyroid cancer was ten times higher than the normal rate for the country.[47]

A 2011 epidemiological review of Chernobyl summarized the impact on millions of people in Europe. The risk of thyroid tumors from exposure to radioactive iodine is greatest for the youngest age groups and perhaps those with an iodine deficiency. Chernobyl workers revealed evidence of increased risk for leukemias and cataracts.

---

[45] World Health Organization, Fact Sheet" Health Effect of the Chernobyl Accident: An Overview", April 2006, www.who.int/mediacentre/factsheets/fs303/en/index.html

[46] NRC, "High Radiation Doses", March 31, 2011, www.nrc.gov/about-nrc/radiation/health-effects/high-rad-doses.html

[47] Shoigu, SK, National Statement, IEAE Updating Report on the International Chernobyl Project Minister for Civil Defense, Emergencies and Elimination of Consequences of Natural Disasters. P53, One Decade After Chernobyl, www.pub.iaea.org/MTCD/Publications/PDF/Pub 1001_web.pdf

Low dose radiation may increased the risk of cardiovascular diseases.[48]

A 2010 study found the risk of leukemia significantly increased in children less than 6 years old at the time of Chernobyl who had a radiation exposure higher than 10mGy.[49]

Doses as low as 250mSv were associated with cataracts in Chernobyl victims.[50] Previously, it was assumed more than 2 Sv radiation, a much higher dose, could induce cataracts.

What happened to the Chernobyl area environment?

The environment around Chernobyl has also been affected by radiation contamination. Cellular and molecular changes have been observed in animals and plants. Conifers were particularly sensitive. The trees received doses of 80 to 100 Gy and experienced large-scale die-off.

Pathological modifications were detected in rodents and invertebrates. [51] There are hot spots of cesium-137 reported of 370,000 kBq/m2 (10000 ci/km2) and strontium-90 hot spots of 185,000 kBq/m2

---

[48] Cardis, E, Hatch, M, "The Chernobyl Accident-An Epidemiological Perspective", Clin Oncol, 2011, May;23 (4) 251-60, www.pub.iaea.org/MTCD/Publications/PDF/Pub 1001_web.pdf, 21396807

[49] Noshchenko, AG, Bodar, OY, Drozdova, VD, "Radiation-Induced Leukemia Among Children Aged 1-5 Years At the Time of the Chernobyl Accident", Int. J. Cancer, 2010, Jul 15;127(2) 412-26, pubmed 19688829

[50] WHO, "Health Effects of the Chernobyl Accident: An Overview", Fact Sheet Media Centre, www.who.int/mediacentre/factsheets/fs303/en/index.html

[51] Robeau,. DG, "One Decade After Chernobyl: Environmental Impact Assessment", IEAE Updating Report p74-76 IAEA Updating Report, www.pub.iaea.org/MTCD/Publications/PDF/Pub 1001_web.pdf

(5000Ci/km2). There are also plutonium hot spots of up to 555kBq/m3.[52]

Direct costs of the accident are estimated at $7 billion.

How did authorities react to Chernobyl victims?

Reports of thyroid cancers and other consequences were initially dismissed based on confounding factors and inconsistent dose-response in the region. Recent Chernobyl reports have usually corrected the premature dismissal of health consequences.

Initially, the increase in thyroid cancer rates was questioned by skeptical investigators because of endemic goiter in the population. The investigators also questioned the pattern of cases because they were not uniform with the dosage increase over the geography of the plume.

The Chernobyl Forum report of 2003-2005 stated, "apart from the dramatic increase in thyroid cancer incidence in those exposed at a young age, there is no clearly demonstrated increase in the incidence of solid cancers or leukemia due to radiation in the affected population." [53] The report added there was an increase in psychological problems.

An UNSCEAR report of 2008 stated, "although those exposed as children and the emergency and recovery workers are at increased risk of radiation-induced effects,

[52] Jensen, P Hedemann, "One Decade After Chernobyl: Environmental Impact Assessments", IAEA Updating Report p82c, www.pub.iaea.org/MTCD/Publications/PDF/Pub 1001_web.pdf

[53] Chernobyl Forum, IAEA, WHO, UN, FAO, UNSCEAR, "Chernobyl's Legacy Health, Environment and Socio Economic Impacts", 2003-2004 p 7 www.iaea.org/publications/booklets/chernobyl/chernobyl.pdf

the vast majority of the population need not live in fear of serious health consequences"[54]

A researcher found that although a substantial increase of thyroid cancer was observed among exposed children, there was no meaningful statistical association to leukemia. The author exempted Russian clean-up workers from his conclusion and proposed more studies[55]. A 2011 study confirmed the leukemia risk.

A WHO April 2006 report suggested significant non-radiation related reduction in average lifespan in the past 15 years in the region. Overuse of tobacco and alcohol and reduced healthcare were considered factors.[56] Smoking and alcohol overuse are equated to a level 7 radiation accident?

Chernobyl investigators also reported,"radiophobia" based on stress and fear of radiation that led to 200,000 abortions of healthy fetuses and alcoholism.

Fukushima Daiichi

On March 11 of 2011, a magnitude 9.0 earthquake and tsunami damaged the power systems at the nuclear power plant, causing the cooling systems to fail. The Tokyo Electric Power Company's Fukushima Daiichi Nuclear Power Plant was shut down. However, the nuclear materials still created heat and had to be kept submerged to cool and prevent melting. The backup generators for

---

[54]UNSCEAR 2008, Report, Vol. II, p 65 www.unscear,org/docs/reports/2008/11-80076-Report_2008_AnnexD.pdf

[55] Howe, GR, "Leukemia Following the Chernobyl Accident", Health Phys. 2007 Nov 93(5):512-5, pubmed 18049227

[56] World Health Organization Expert Group, "Health Effects of the Chernobyl Accident: An Overview" April 2006,

the water pumps to cool the radioactive material ran out of fuel during the outage. The material over heated.

On March 12, radioactive steam was released into the air to promote the cooling process. Furthermore, on March 12 a hydrogen reaction occurred. Hydrogen is a byproduct of the cooling process. Radioactive cesium, iodine, and strontium were released.

By March 14, a total of 185,000 residents had been evacuated away from the power plant. According to the IAEA, 230,000 units of stable iodine had been distributed to evacuation centers, for use if needed. Soon, all three suppliers of potassium iodide were out of stock.

Radiation levels on March 15 increased to 400 millisieverts per hour. Radiation was detected in California and Hawaii. A week later, radiation from the accident was reported in EPA monitors in many US States and eventually in Germany.

Officials evacuated residents living within 20 kms of the plant and recommended voluntary evacuation for those living 20 kms to 30 kms from the plant. [57]

On March 19, the Japanese public health authorities told the water utilities to stop intake from surface water sources and cover the water-treatment plants with plastic sheets. From March 21 to April 1 the Fukushima water supply was halted because radiation exceeded the standards.

The U.S. EPA released radiation monitoring information on their website during the event from March to June.

---

[57] IAEA, Incident and Emergency Centre, "Status of the Fukushima Daiichi, Nuclear Power Plants"

Normally the radiation monitoring is reported in quarterly summaries. The U.S. EPA RadNet radiation monitoring program reported;

- Both Cesium 134 and 137 in air filters on April 4 in Juneau and Nome Alaska, Anaheim, San Bernadino and San Francisco, California, Jacksonville and Orlando Florida, Honolulu and Kauai and Oahu Hawaii, Boise Idaho, Las Vegas Nevada and Salt Lake City Utah.

- Uranium 238 was found in Riverside California and Kauai Hawaii.

Food samples tested in March to early April in Fukushima revealed 13% of the milk and 14% of the produce exceeded the limits for radioactive iodine. 11% of the produce also exceeded the limits for radioactive cesium. Samples in late May recorded 52 of 818 with cesium or iodine radiation above legal limits.[58]

On March 29 the water discharges from the Fukushima plant had 3,355 times the legal limit of iodine. In sampling of April 3, sea water contamination was 37.5 Bq/L, about half of the level of March 30. The soil contamination was 21 Bq/kg, in the range of 20-50 since March 25. On March 21 and 22 plutonium was found in soil samples at the plant.[59]

There are reportedly, many nuclear power plants in the U.S. with the same design as the Fukushima plant. The accident would not have occurred if the cooling water

---

[58]WHO, Western Pacific Region, Japanese Earthquake Situation Report, March 15 2011 report www.wpro.who.int/nr/rdonlyres/DE6CD789-ED92-4590

[59]WHO, Western Pacific Region, "SitRep#25" Japanese Earthquake and Tsunami, April 6 2011, www.wpro.who.int/nr/rdonlyres/DE6CD789-ED92-4590

flowed by gravity instead of pumps. Could any experienced plumber have predicted the outcome and designed the solution?

Have there been other accidents or incidents?

On March 28, 1979 the Three Mile Island nuclear reactor fuel heated and partially melted causing a release of a small amount of radioactive iodine and xenon. The accident was due to a combination of equipment malfunction and then operator errors.

In the U.S., there are 55,000 events described in licensee reports to the NRC where a personnel error, power loss, storm, or equipment malfunction created an event, and a potential for a nuclear incident or accident. The list of events is searchable at https://lersearch.nil.gov. Most events were not dramatic, but some sound like the combination of operator errors and equipment malfunctions that have led to major accidents.

The following is a random sampling of the events reported at the NRC licensee event reporting site.[60]

- On May 7 1994, at the Dresden 1 nuclear power plant, 50,000 gallons of contaminated water were released to ground water.

- In December 1985, at the Yankee Atomic Electric Company, a contractor who was vacuuming inside the main control board accidentally bumped against the reactor protection system auxiliary circuit relay with a cleaning attachment. A reactor scram

---

[60] U.S. NRC, Licensee Event Report Search, www.lersearch.inl.gov

occurred. (Reactor scram is an emergency shut-down to prevent an accident)

- On June 16 1991, a lightning strike destroyed a lightning arrestor. All power was lost and a reactor scram occurred. The States of Vermont and Massachusetts were notified.

- On November 15, 2000 at the Oyster Creek plant, the report title states "reactor scram due to low reactor water level resulting from personnel error."

- On October 2, 2000 at the Nine Mile Pt. 1 nuclear plant a manual reactor scram and unusual event declaration were due to a stuck open electromagnetic relief valve and failed vacuum breaker.

- On May 4 2007 at the Dresden 2 plant, a reactor scram occurred due to loss of feed water.

- On April 23, 1989 at the Yankee Rowe plant, dropped control rods resulted in a reactor scram on low main coolant pressure.

In March 2011, the Nuclear Regulatory Commission rejected challenges to extending an operating license for another 20 years to the Vermont Yankee nuclear plant. The State of Vermont had argued the plant was too old and unreliable to continue. The plant had already operated on the original 40 year license. There are currently 38 nuclear power plants with applications for re-licensing in process.

Anyone living within fifty miles of a nuclear plant should have an emergency plan, an emergency radiation kit and shelter with provisions. They should also know the emergency plans of the community for a radiation event.

Are the airport body screenings potentially dangerous?

It is a small dose if the machines are functioning correctly. The backscatter x-ray machines at airports are subjecting fliers to a 0.005 to 0.009 mrem dose. The scan takes at least eight seconds.

The NCRP, (National Council on Radiation Protection and Measurements), has defined a "negligible dose" as 1 mrem per year. If a flier travels once per week for a year, they will exceed the negligible dose. Authorities state that the public security benefit is calculated against the minute risk of cancer per million flyers.

There are a few issues that have been debated among scientists recently;

- The government states that the machines will be inspected initially, then annually. However, these machines are in use all day, every day. Similar machines in occasional use in hospitals are inspected every day.

- If a machine malfunctions, then the screening staff must recognize the problem and call for repair. In a hospital setting, operators receive years of advanced training on the equipment.

- The head is included in the scan, exposing the sensitive lens of the eye. The cornea is not capable of repair.

- The skin receives a high dose of an intense x ray beam. The radiation is reportedly half depleted at 4 cm depth. In adults it would not reach organs, but in children it could.

- A portion of the population is more susceptible to radiation. These include the elderly, those with DNA mutations of cell repair, those with a history of cancer, children, pregnant women and adolescents.

- The scanners are open on the sides and top, potentially exposing the workers stationed nearby.

- The machines are mechanical. A malfunction could allow excessive exposure.

- There has been no independent safety testing.

The original presidential report on using radiation for security also proposed additional "limited use" systems that would be used with discretion on some individuals. These systems could produce a dose where the limit would be reached at only 25 scans. [61] The report states that users of the system would be responsible for keeping track of the total doses received. The persons being scanned are responsible for monitoring their dose?

The same report suggested future uses of scanners in banks, government buildings, county courthouses and other locations.

---

[61]National Council on Radiation Protection and Measurements, "Presidential Report on Radiation Protection Advice: Screening of Humans for Security Purposes Using Ionizing Radiation Scanning Systems, p 10
www.fda.gov/ohrms/dockets/ac/03/briefing/3987b1_pres-report.pdf

Chapter Four

Nuclear Waste and Accidents

How are radioactive materials transported?

Radioactive materials are transported by highways, rail, air, and sea. Three million packages in restricted quantities have been shipped. The NRC and the Department of Transportation have created a warning system based on the degree of radioactivity present. If the symbol on the package has a white background, it has a maximum dose rate of .5mrem/hr and denotes low to moderate level radiation.

The white symbol packaging is not necessarily accident proof during transit, but the potential contamination is limited. These radioactive containers can be transported with other goods. The packaging for these less radioactive containers has failed in 10% of the accidents.

If the radiation symbol background is yellow, it represents a maximum dose of over 50 mrem/hr or even 200mrem/hr if there is yellow placarding of the vehicle in addition to the package. The more highly radioactive products are packaged to avoid disbursal. There has been only one documented failure of packaging during an accident.[62]

---

[62]US NRC "Transportation of Radioactive Material",www.mitnse.files.wordpress.com/2011/03/transportatiion_11.pdf

The warning signs for radioactive dangers are different when not in transport.

How are radioactive materials usually labeled?

A commonly used high radiation area caution sign has a white background.

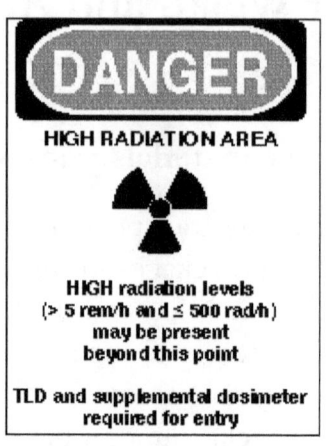

This is the basic trifoil sign as the international symbol of radiation with a yellow background and magenta image.

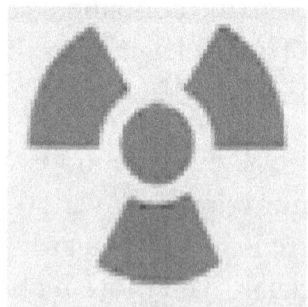

The following is the new supplemental ionizing radiation warning symbol presented by the International Organization for Standardization and the IAEA. It has a red background and black image.

How is radioactive waste classified?

The Nuclear Regulatory Commission sets standards for low-level radioactive waste as Classes, A. B and C. There are currently three of these waste sites in the U.S., Barnwell, SC, Richland, WA, and Clive, Utah. [63] The IAEA classifies low-level waste as either short lived (less than 30 years half life) or long lived (>30 years half life).

Where is nuclear waste disposed?

Short lived low-level waste is currently disposed of in near surface disposal. Forty locations around the world are expected to be approved by the IAEA. Between 1946 and 1982 this waste was dumped in the ocean in metal 55-gallon drums lined with concrete. [64] Strontium-90, cesium-137, cobalt-60 and tritium were disposed in this manner. Known dump locations began in the Pacific

---

[63] U.S. NRC, "Low-Level Waste Disposal", May 3, 2011, www.nrc.gov/waste/llw-disposal.html

[64] IAEA Bulletin, "Ocean Disposal of Radioactive Waste", 4/1989, www.iaea.org/publications/magazines/bulletins/bull314/31404684750.pdf

Ocean 80 kms from California and ended in the Atlantic Ocean 550 kms from the European shelf. The dumping operations were under the control of the Nuclear Energy Agency of the Organisation for Economic Cooperation and Development. In 1982, a voluntary moratorium was placed on ocean dumping.

Long lived low-level waste is stored in geologic repositories.

High-Level Waste is a separate category. This waste must be stored away from the reactor for 5 to 100 years and then be reprocessed.

Have there been any transportation accidents involving nuclear materials?

On January 16 1966, a B52 aircraft carrying nuclear weapons crashed in Palomares, Spain. Over 600 acres were contaminated with plutonium. Almost 5000 metal 55-gallon drums of contaminated soil were shipped to Savannah Naval Storage in Aiken, South Carolina.

Spain was reportedly concerned about public panic and never restricted the area. Of 714 people followed until 1988, 124 still showed plutonium in their urine.[65]

In January 1968 a U.S. Air Force plane with nuclear weapons crash landed in Thule, Greenland. The plutonium burned and disbursed as particles. The plutonium was later found in shellfish at concentrations a thousand times higher than normal. 24,000 tons of

---

[65] Agency for Toxic Substances and Disease Registry, "Radiation Accidents", Ionizing Radiation p 195

contaminated ice and snow were excavated and shipped to the U.S. [66]

Have there been military weapon production facility accidents?

In 1957 in Kyshtym, a Soviet Union plutonium production facility for the military had a cooling system malfunction, and the waste storage tanks exploded. Cesium, zirconium, niobium and strontium were released. 10,000 people were evacuated from highly contaminated areas and 260,000 people remained in a less radioactive area.

A Soviet researcher revealed that if a medical exam found evidence of a radiation disease, physicians were not allowed to relate it to radiation. The researcher states there are no Russian reports of the impact of nuclear plants on health.

In October 1957, radioactive material was released from the Windscale, UK nuclear weapons plant at Sellafield, UK. An operator error during a routine release of energy caused the fuel to overheat. Uranium was oxidized, and a fire started. The fire burned for three days. Iodine, cesium, polonium, ruthenium and xenon were released. Contamination of pastureland was extensive. [67] It is estimated children living nearby were exposed to 100 mGy radioactive iodine.

---

[66] Agency for Toxic Substances and Disease Registry, "Radiation Accidents", Ionizing Radiation p 199

[67] Agency for Toxic Substances and Disease Registry, "Radiation Accidents", Ionizing Radiation p 208

A 10-fold increased incidence of leukemia and non-Hodgkins lymphoma in children near Sellafield has been reported. Researchers have hypothesized about the cause of the illnesses of children born in the local area. The investigators report the association with fathers exposed to radiation before conception is not clear. [68]

One Sellafield researcher reported that the paternal irradiation theory did not have support and believed an infectious agent as the cause because of population and socioeconomic mixing.[69] Do they believe that the 1957 accident is not relevant?

A study of Sellafield workers found a significant excess of deaths from cancer of the breast and secondary cancers. There were trends in lymphatic and hematopoietic cancers. The researchers concluded there was not an overall significantly increased risk of cancer compared to other radiation workers. [70]

Have there been any accidents involving medical equipment?

In September 1987, a piece of medical radiology equipment was scavenged from an abandoned clinic in Goiania Brazil. Cesium-137 was present in the equipment that the junk dealer broke open. 249 people were eventually contaminated externally or internally through

[68] Gardner, MJ, "Leukemia in Children and Paternal Radiation Exposure at the Sellafield Nuclear Site," J. Natl. Cancer Inst. Monogr. 1992;(12) 133-5, 1616797

[69] Kinlen LJ, "Childhood Leukemia and Non-Hodgkins Lymphoma in Young People Living Close to Nuclear Reprocessing Sites", Biomed Pharmacother.1993;47(10);429-34pubmed/8061241

[70] Omar, RZ, Barber, JA, Smith, PG, "Cancer Mortality and Morbidity Among Plutonium Workers at the Sellafield Plant of British Nuclear Fuels", Br. J. Cancer, 1999 Mar; 79(7-8) 1288-301, pubmed 10098774

ingestion of contaminated fruits and vegetables. Four victims died. [71]

In 1983, a medical therapy machine containing cobalt was sold as scrap metal in Ciudad Juarez, Mexico. Thousands of tons of steel sold in Mexico and the U.S. were contaminated. It is estimated a thousand people were exposed without knowledge.

What regulations control water contamination?

Radiation contaminates ground water through natural decay of the rocks and through downward migration of surface contamination. Atmospheric radiation easily contaminates surface water sources, lakes and streams. The EPA sets drinking water contaminant limits for radionuclides. The goal for radioactive contaminants is zero. However, the maximum contaminant limit is;[72]

Alpha particles------15 pCi/L

Beta particles--------4 mrem/year

Uranium -------------30 ug/L

Tritium --------------20,000 pCi/L (4 mrem/year)

---

[71] Agency for Toxic Substances and Disease Registry, "Radiation Accidents", Ionizing Radiation p 197, www.atsdr.cdc.gov/toxprofiles

[72] US EPA, "Drinking Water Contaminants, http://water.epa.gov/drink/contaminants/index.cfm

Chapter Five

Nuclear Weapon Testing and Production

When did the nuclear testing of weapons occur?

There were two peaks of fallout radiation from testing, the first during 1954-1959 and the second, 1961 to 1962. Carbon-14, cesium-137, zirconium-95 and strontium-90 were the principal materials released.

Atmospheric nuclear weapon testing produced a significant quantity of airborne radioactive particles. The weapons testing produced strontium-90 for the first time in the planet's history. Over 16 million curies of strontium-90 were produced and globally dispersed from the weapons tests.

The weapon testing started in 1945. The worldwide average ingested radiation dose of Strontium 90 from the atmospheric tests is 9.7 mrem according to the NRC. An additional .92 mrem is inhaled contamination. [73]

A test in the Pacific actually used a 15 Mt detonation. This is equal to the explosive potential of 15,000,000 tons of TNT. The "Castle Bravo" detonation by the U.S. in

---

[73] US NRC, "Backgrounder on Radiation Protection and the Tooth Fairy Issue", Dec. 2004, www.nrc.gov/reading-rm/doc-collections/factsheets/tooth-fairy.pdf

1954 was three times larger than expected because of an error in the lithium content. Traces of radioactivity were found in Australia, India and Japan. One island is still uninhabitable. There were third-degree burns reported 62 miles away. The Soviet Union countered with a 50 Mt "Tsar Bomba"detonated in northern Russia.

The following is a chronological list of some of the documented nuclear weapons tests.

1956 About 80 nuclear tests conducted by the US., Great Britain and the Soviet Union

1956-1958 180 nuclear tests by the U.S., Soviet Union and Great Britain

1960 France conducts 3 nuclear tests

1961-1962 100 nuclear detonations by the US and Soviet Union

1963 The U.S. and Soviet Union cease nuclear tests.

1964 China conducts first nuclear detonation

1965-1967 Nine nuclear tests by France and China

1968-1970 Three nuclear tests by China

1971-1974 24 nuclear detonations by France and China

1976-1978 Six nuclear tests conducted by China

How could so many countries develop nuclear technology?

The proliferation of the nuclear weapon testing was encouraged by the theft of nuclear technology by spies. Klaus Fuchs, a German refugee and assistant at Los Alamos for the Manhattan Project, spied for the Soviets while working on the Manhattan Project since 1941. He was convicted and sentenced to 14 years in prison.

Pakistan obtained nuclear technology through the efforts of a nuclear scientist who took blueprints from his employer, a uranium processing company in the Netherlands in 1974. Abdul Qadir Khan's Pakistani nuclear program was known but ignored by the US, and other nations. In January, 2004 Kahn confessed that he ran a supply network of nuclear weapon secrets to Iran, Libya and North Korea. He was placed under house arrest in Pakistan but released in 2009.

What countries tested the most weapons?

The following is the number of detonations by country;

United States, 1945-1992 1054 detonations,

Russia, 1949-1990 715 detonations

France, 210 tests (50 atmospheric, 160 underground)

United Kingdom, 45 tests

China, 45 tests (23 atmospheric, 22 underground)

India, 1974-1998 6 detonations

Pakistan, 1998 6 detonations

North Korea, 2006-2009 2 detonations (underground)

Are underground tests safer?

In addition to atmospheric tests, there have been about 1400 underground nuclear tests worldwide with a total yield of 90 Mt explosives. Some radioactive material can be released to the atmosphere if the blast penetrates the surface. Of 500 underground tests in Nevada, 32 led to off-site contamination.[74]

Have nuclear weapons been stolen?

The IAEA reports hundreds of unauthorized possession to the agency's Illicit Trafficking Database. [75]

The IEAE reports that in the year, July 2009 to June 2010, there were 222 incidents reported to the Illicit Trafficking Database. 61 incidents involved theft and five involved highly enriched uranium or plutonium.[76] In the prior sixteen years, 1993 to 2009 there had been a total of 1773 incidents, but only 351 involved unauthorized possession and criminal activities. The issue of illicit trafficking is still a major concern.

Where were the weapons produced?

The manufacturing of nuclear weapons required a number of production facilities in the U. S. Due to the long half-life of the radioactive waste, the facilities and their inventories are still a potential health hazard.

---

[74] ATSDR, "Sources of Population Exposure to Ionizing Radiation", p 269

[75] Dr. Patel, Gordhan N., "Radiation Dosimeter and Dirty Bomb", Testimony before the U.S. House Committee on Government Reform, Subcommittee on National Security, Emerging Threats and International Relations on Counterterrorism Technology, September 29, 2003 p3

[76]IAEA, "Illicit Trafficking Database", ITDB, www.ns.iaea.org/security/itdb.asp

The Fernald uranium production center in Ohio was called the "Feed Materials Production Center,." In 37 years of operation, 31 million pounds of uranium were produced. There were 2.5 billion pounds of hazardous waste. A 223 acre area of the local water aquifer is reportedly contaminated with uranium.

At Oak Ridge National Laboratories, 200 acres are contaminated in eastern Tennessee. Uranium enrichment was conducted until 1985.

At Paducah, Kentucky fifty years of enrichment of uranium for nuclear reactor fuel contaminated 3,400 acres in Western Kentucky.

At Portsmouth Ohio, enrichment processes for technetium, cesium, strontium, plutonium and neptunium, continued from 1952 to 2001.

At Hanford, Washington the WWII facility produced plutonium for weapons until the 1960's. There are 1.7 trillion liters of liquid waste and 625,000 meters of solid waste.

At the Rocky Flats CO manufacturing facility near Denver, plutonium, beryllium and uranium were manufactured for weapons from 1951 to 1991.

At the Idaho National Laboratory in southeast Idaho, 52 nuclear reactors have reprocessed nuclear fuel since 1949. There are 284 tons of spent nuclear fuel stored at the location.

During the 1950's nuclear material production facilities in the Soviet Union in the South Ural mountains released major radioactive contaminants. Villagers downstream from the plant experienced increased leukemia rates, and workers experienced plutonium-related lung cancers. [77]

Have there been accidents at weapons plants?

In Rocky Flats, Colorado a nuclear weapons plant experienced fires in 1957 and in 1969 when the plutonium spontaneously ignited. The smoke plume containing plutonium spread to populated areas near Denver. Although follow-up studies of cancer found some increases, they were determined "not conclusive."

In 1993, at a plutonium production reactor near Tomsk, Russia, a tank exploded, causing the release of uranium, plutonium and other radioactive elements. Gamma radiation was 20 times higher than normal.

Are there plans to decontaminate production facilities?

The Department of Energy, Environmental Management, published a Five Year Plan (2008-2012) for major sites and closed sites. The following are some of the decontamination activities scheduled;

- Stabilizing and packaging plutonium residues at Rocky Flats and Savannah River sites.

- Producing over 2,000 cans of vitrified high-level waste from highly radioactive liquid waste at Savannah River Site and West Valley, NY.

---

[77] Goldman, M, "The Russian Radiation Legacy: Its Integrated Impact and Lessons", Environ Health Perspect. 1997 Dec;105 Suppl6 ; 1385-91/pubmed 9467049

- Retrieving and packaging 2,100 tons of spent nuclear fuel from the Hanford site

- Certifying and shipping 40,000 cubic meters of transuranic waste to a waste isolation Pilot Plant

- Cleaning up Melton Valley at Oak Ridge Reservation

- Disposing of 8,500 tons of scrap metal at Portsmouth Ohio

- Removing mill tailings pile at Moab site in Utah. 16 million tons of Uranium are to be relocated.

- Increasing clean up at Los Alamos National Lab

Chapter Six

––––––

Nuclear Weapon Attack

Is nuclear war still a threat?

Since the Cold War, Americans have been concerned about nuclear attacks. There was such fear of radiation five decades ago, that it was common to see bomb shelters in backyards of subdivisions. Many public buildings displayed nuclear bomb shelter symbols. Announcements on tv and radio addressed radiation safety measures. School children had nuclear war drills.

The scenario for a nuclear threat has changed since the Cold War when a global war between the former Soviet Union and the United States was the major concern. A State sponsored attack by a rogue dictator, serious diplomatic arguments or military coup is still possible, but a limited attack by a terrorist group is more likely.

The following are more likely threats to be considered,

1. A dirty bomb is a conventional explosive device with radioactive contents. This is the easiest type of nuclear attack for a terrorist group to execute. A dirty bomb will not create an atomic blast. It uses dynamite to scatter radioactive dust, smoke or other material to cause radioactive contamination in a limited area. The major acute danger is expected to be from blast injuries. There may be

insufficient radiation to cause acute radiation illness. The radioactive material may be stolen by a terrorist group and detonated at ground level in a major US city. Since this attack does not require a rocket to deliver the weapon, it is a feasible terrorist weapon.

2. Radiological Exposure Devices (RED) are silent sources of radiation with no explosive component. The radioactive device could contaminate thousands before discovery.

3. A limited nuclear war elsewhere in the world could result in atmospheric contamination across the US and possibly affect climate for months or years. A look at the unstable leadership in many of the new nuclear armed countries reveals the risk.

4. A limited nuclear attack against the United States could be caused by an accidental launch, a terrorist group gaining control of a nuclear weapon or an attack by a foreign dictator.

5. A full-scale attack from a major nuclear power.

Are there expected targets?

Potential targets may include geographic regions with strategic missile sites and military bases, capitals of government, transportation and communication centers, manufacturing, technology and financial centers. Also, petroleum refineries, electrical or nuclear power plants, important national symbols and major shipping ports and airports.

If you live within fifty miles of these possible targets, you should develop a personal radiation emergency plan and have a radiation emergency kit with supplies for decontamination and neutralization. Weather conditions during fallout could expand the geographic distance.

What would a nuclear detonation look like?

The first sign is a bright flash of light. A nuclear blast is created by either fission or fusion of atoms that produces an intense pulse or wave of heat, light, pressure and radiation. A large fireball is created, and all material inside the fireball vaporizes and is carried upwards. This is the mushroom cloud. In the mushroom cloud, the radioactive elements mix with the vaporized debris. This is fallout. As it cools, it falls back to earth, carried with the wind and blast. Fallout increases if there is rain or snow. In the first minute after detonation, highly penetrating gamma and neutron radiation is emitted

Nuclear weapons are measured based on the equivalent conventional explosives. A 10 Kt nuclear weapon is equal to 10,000 tons of TNT. A 10 Mt weapon is equal to 10,000,000 tons of TNT.

The heat and wind effects will be lethal for anyone exposed without shelter. A 10 Kt effect can cause first degree burns two miles away from the blast. A shockwave of pressure moving away from the detonation can create air pressure winds of 160 mph a quarter-mile from the blast within seconds. Acute injuries of burns and cuts will be the focus of medical attention. Victims should avoid windows and seek interior rooms or underground protection.

The initial impact and deadly force of a nuclear bomb is not the radiation, but the heat and wind. The pressure wave from a 1 Mt atomic bomb would destroy all buildings and 98% of the population within two miles of the bomb blast. Most homes within a 5-mile radius would be destroyed. A 20 Mt bomb would destroy homes within a 16-mile radius. Death from fallout radiation would extend many miles farther. Winds would carry the fallout radiation further yet. Fortunately, a nuclear attack will probably not be this large. Federal planners base their disaster planning on a 10 Kt weapon.

The extent of the radiation contamination will be based on the size of the detonation, height above the ground, (closer to the ground creates more fallout), the surface beneath the explosion, (flat areas are more susceptible to blast), and weather conditions.

Will there be a warning of a nuclear detonation?

The National Warning System (NAWAS) is designed to disseminate information about dangerous weather or terrorist emergencies. A warning would be given by sirens, but the meaning would not be recognized by most people. The Emergency Broadcast System (EBS) uses selected AM, FM, and TV stations to transmit messages.

The problem with this electronic warning, is that communication by TV or radio will probably not be possible after the detonation. Any nuclear detonation is accompanied by an EMP (Electromagnetic Pulse) that destroys all electronics. The larger the detonation, the wider the area affected. Only electronics stored in a Faraday cage can be protected from the EMP and continue to function after the nuclear detonation. We encourage readers to read <u>EMP Survival: How to Prepare</u> <u>Now to Survive When an Electromagnetic Pulse Destroys</u> <u>Our Power Grid.</u>

How should a person respond to a nuclear detonation?

At the time of detonation, gamma radiation, which travels at the speed of light will reach you instantly. The only thing to protect against gamma radiation is heavy materials such as lead, steel, dirt, or concrete.

If you notice a bright flash and the power goes out find shelter immediately. The flash is brighter than sunlight. It is so bright it can cause temporary blindness for 30 minutes and possible long-term damage to the eye.

You will have a few seconds before the blast waves of heat and wind. The blast of heat will last from 8 seconds to 44 seconds based on the size of the blast. If you are driving, find shelter in a building or underground. If you are outside, lay down inside of a deep culvert or concrete tunnel if possible. Find the best concrete or brick shield. Just stepping inside a building can reduce your exposure to a tenth of the outside level. Your radiation exposure will be about ten times less in the basement than upstairs. Drop to the ground face down in your best shelter and place your hands over your mouth and nose. Remain covered and protected for at least two minutes awaiting the wind and heat blasts.

You are much more exposed standing up than lying down. Remain flat until the heat and two shock waves pass. The initial waves from the detonation will cause blast (lacerations, punctures) and thermal (burn) injuries especially if you are near windows or unshielded.

If the blasts do not reach you in two minutes then it is over 25 miles away.

What happens after the initial blast waves?

After the waves of heat and wind, victims must continue to remain adequately sheltered until the radioactive fallout level is safe. You will need a radiation detector to determine when it is safe to leave. It could require hours, days, or even weeks.

In later chapters details will be provided in finding the best adequate shelter, decontaminating yourself and your shelter and preventing further exposure from food, water and air.

Chapter Seven

_____

Radiation Neutralizers

This chapter provides information on the treatments and research on potential neutralizers we discovered in the literature. We advise readers to continue to monitor the status of research into radionuclide neutralizers.

Many treatments are specific to blocking a single radionuclide. The primary radioactive elements in nuclear detonations and nuclear accidents are iodine-131, strontium-90, cesium-134 and cesium-137. Strontium settles near the disaster site, and cesium and iodine are distributed much farther by the wind. Iodine will decay within 90 days. Cesium-137 and strontium-90 will last decades.

There are radiation decorporators (treatments to reduce the effect of radiation) available by prescription and in the National Stockpile of disaster supplies. The radiation decorporators that may be used by physicians are described. Later in the chapter we present information on current animal research of other potential radiation neutralizers. These non prescription options should be discussed with your physician and added to your supply if your physician approves.

We advise you to seek medical care and physician's instructions for any radiation treatment if possible.

What would you expect for treatment if medical assistance is available?

Prussian blue (ferric hexacynoferrate)

Prussian blue is a pigment that removes radioactive cesium from the intestine. Since the 1960's, it has been used to treat contamination by cesium and thallium. It is commonly used as a laboratory stain in pathology labs. The prussian blue traps cesium and thallium in the intestines and keeps it from being re-absorbed into the body. The radioactive materials are then excreted in bowel movements. Prussian blue was first used as a blue dye in 1704 for artists and manufacturers.

Prussian blue is available by prescription for radiation treatment. Radiogardase® is the name of the product approved by the FDA for cesium and thallium exposure. The recommended dose is 3 grams orally three times a day for adults.[78] Check the prescribing information for details on pediatric dosing.

The International Atomic Energy Agency (IAEA) recommends an adult take 10 grams a day. Prussian blue is still effective after time has elapsed after exposure. Treatment should continue for 30 days. When the radioactivity is substantially decreased, dosing can be decreased to 1 to 2 grams. Bioassay of feces and urine samples at a hospital can measure cesium contamination.

It is included in the strategic National Stockpile for radiation emergencies.

---

[78] Radiogardase, Prussian blue insoluble capsules, prescribing information, Fdswa150\nonectd\N21626\S_007\2008-04-21\spl\radiogardaase.xml

Belarus, Russia and Ukraine settlements used prussian blue after Chernobyl as an additive in animal food. Two to eightfold reductions in radioactive cesium were observed in milk and meat from cattle grazing on contaminated lands and treated with Prussian blue. [79]

One study suggested that use of prussian blue (ferrocin)-containing food in populated areas of strict radiation control could permit considerable reduction of the internal dose of radioactive cesium.[80]

Prussian blue is the preferred treatment for some radionuclides, including; cesium, rubidium, thallium according to the NCRP (National Council on Radiation Protection and Measurements) "Decorporation Therapy Recommendations". [81]

Ca-DTPA and Zn- DTPA

DTPA is available by prescription only. It stands for diethylenetriaminepentaacetate, with zinc or calcium. DTPA is a chelating agent that binds and holds onto radioactive materials in the body. The chelating agent then is excreted through the urine. It is approved currently for only three radioactive elements; plutonium, americium and curium. It should be given the first day of exposure. On the first day Ca-DTPA is ten times more effective than Zn-DTPA. After 24 hours, both are equally

[79] Richards, JI ,et al "One Decade After Chernobyl, the FAO Response,", Keynote Presentation, IAEA Updating International Chernobyl Project, p 137 www.-pub.iaea.org/MTCD/publications/PDF/Pub1001_web.pdf

[80] Korzun, VN, "Decrease In Internal Irradiation Dose From Cesium Radionuclides by Using Ferrocin", (Prussian Blue) Med. Radiol. 1991, 36 (5) 23-7 www.pubmed/2034101

[81]NCRP Report 161, 2010, Radiation Emergency Assistance Center REAC/TS "Medical Aspects of Radiation Incidents", www.orise.orau.gov/reacts p29-30

effective. DTPA will continue to work, although less effectively, for days or weeks.

DTPA is available by injection or IV. In past experience, most people only needed a single dose of DTPA.

Patients who inhaled radioactive elements can be administered DTPA in a mist or spray. Some medical authorities recommend DTPA in a nebulizer for inhalation contamination of the lungs.

According to the NCRP, "Decorporation Therapy Recommendations" DTPA is the recommended treatment for cobalt, yttrium, americium, berkelium, californium, cerium, chromium, cobalt, curium, einsteinium, europium, indium, and iridium, in addition to the approved radionuclides previously mentioned.[82]

DTPA should not be used to treat people internally contaminated with uranium, or neptunium. DTPA may remove important minerals like zinc, magnesium and manganese. [83]

Potassium Iodide (KI)

If radioactive iodine is absorbed internally, the thyroid gland quickly absorbs it where it causes injury. Potassium iodide is a salt of stable (not radioactive) iodine. The potassium iodide blocks the radioactive iodine from being absorbed by the thyroid. Potassium iodide can only protect the thyroid, not other parts of the body. It can block radioactive iodine, but not other radioactive

---

[82]Radiation Emergency Assistance Center, "The Medical Aspects of Radiation Incidents",Table 7, Decorporation Therapy Recommendation in the USA for Radionuclides of Concern, NCRP Report 161 2010 p 29-31

[83] Center for Disease Control and Prevention, Emergency Preparedness and Response, Facts about DTPA, www.bt.cdc.gov/radiation/dtpa.asp

materials. In case you are wondering, table salt with iodine is insufficient to block radioactive iodine.

The effectiveness depends on;

- the time between contamination and taking potassium iodide,

- the absorption of potassium iodide, and

- the total amount of radioactive iodine exposure.

Based on Chernobyl experience, young children and those with low levels of iodine are at highest risk of injury. The most effective timing of treatment is within the first three hours of exposure.

Iosat™, ThyroSafe™ and ThyroShield™ Solution are FDA approved brand names of potassium iodide. The FDA has approved tablets and liquid form of potassium iodide in 130 mg and 65 mg. Adults should take 130 mg; children 3 to 18 years should take 65 mg. Read the package directions for detailed instructions. A single dose protects the thyroid for 24 hours. A one-time dose is usually sufficient. Potassium iodide is available without a prescription currently.

For further information read FDA recommendations on KI (Potassium Iodide) www.fda.gov/Drugs/EmergencyPrepardness/Bioterroris mandDrug Preparedness/ucm072265.htm.

80% of the Chernobyl accident survivors report they did not receive potassium iodide. Fukushima evacuees may not have been given the potassium iodide promptly based on the update days after the event. The suppliers temporarily ran out of stock soon after Fukushima.

Potassium iodide is currently available without a prescription from internet suppliers. It should be in your personal radiation emergency kit. It is best to wait for physicians instructions before taking potassium iodide. Some people are allergic to iodine or have thyroid disorders and should not take additional iodine.

Calcium Chloride Suspension or Calcium gluconate

Radioactive radium and strontium treatments include 10% calcium chloride suspension by IV or calcium gluconate or aluminum hydroxide by mouth. These treatments are prescribed and administered by physicians. Calcium chloride is only available by prescription.

Water Diuresis

Diuresis is useful to increase urinary excretion of Tritium. A physician will determine if your radiation exposure and your medical condition would suggest this therapy.

Sodium Bicarbonate

A study of dogs long ago revealed that uranium nephritis and acidosis were diminished by the administration of oral sodium bicarbonate.[84] It is currently included in the list of medical treatments for uranium only as oral or IV administration. This therapy is administered by physicians

---

[84]Goto, Kingo, "A Study of the Acidosis Blood Urea, and Plasma Chlorides in Uranium Nephritis in the Dog and of the Protective Action of Sodium Bicarbonate, Jrnl Exper. Med, JEM, Vol 25 #5, May 1 1917, 693

What neutralizers may your physician approve for your emergency kit, if you anticipate no medical assistance?

In a radiation emergency, the treatments available from physicians, described in the previous section are more thoroughly tested and proven safe and effective.

Research into radionuclide neutralizers is not complete or conclusive. Most of the research is based on animal studies. Some neutralizers were used in practice after Chernobyl. Others have not been tested. Most are available without a prescription as nutritional supplements. We encourage readers to continue to follow the research in the use and dosing of these products for radiation injury.

We also encourage readers to ask their physicians about these supplements. Based upon your further research and consultation with your physician you could include these potential neutralizers in your emergency radiation kit. Based on your consultation with your physician and your blood tests, you may be advised to take some of these supplements.

If you are in a shelter waiting for medical attention, you can perform first aid, and clean and protect wounds from further contamination. It is important to wait for radiation treatment until it is known what type of radiation, if any, you were exposed to.

Kelp

Based on the experience at Chernobyl, it is known that those with low iodine levels are at increased risk of radioactive iodine absorption. Most Americans attain a sufficient iodine level with iodized salt. However, some sources report 15% of Americans have a moderate to severe deficiency in iodine. They may not be using the iodized salt, or they may have cut back their salt intake entirely. If your physician determines that you are deficient in iodine, kelp tablets are a good source of replenishment and may be recommended. Some people are allergic to iodine or have thyroid conditions that prevent supplementation.

Fucoidan

Fucoidan is a radiation detoxifier according to animal research. Bladderwrack is an iodine-rich seaweed that contains fucoidan.

Animal studies support the benefit of fucoidan;

- In one animal study, mice pre-treated with doses of fucoidan before irradiation and monitored for 30 days showed significant improvement in survival at 100 mg/kg body weight dose. The authors theorize it was due to the hematopoietic cell (blood producing) viability, possibly through antioxidant and anti-inflammatory mechanisms. [85]

- In a separate animal study, fucoidan was found to have radioprotective effects on bone marrow cells, in cell viability and immuno-activity. [86]

---

[85] Lee, J, Kim, J, Moon C, et al, "Radioprotective Effects of Fucoidan in Mice Treated with Total Body Irradiation", Phytother. Res. 2008, Dec; 22, (12) 1677-81 pubmed/18683851

[86] Byon, YY, Kim, MH, Yoo, ES, Hwang, KK, et al, "Radioprotective Effects of Fucoidan on Bone Marrow Cells: Improvement of the Cell Survival and Immunoreactivity", J. Vet Sci, 2008- Dec 9 (4) 359-65 pubmed/19043310

Fucoidan also has been found to inhibit metastasis by preventing adhesion of tumor cells. [87] Fucoidan extract is available from a variety of nutrition supplement providers.

Sodium Alginate

Sodium alginate efficiently binds strontium-90 in the gastrointestinal tract and prevents absorption according to animal research. Sodium alginate is a constituent of kelp and other sea vegetables. Capsules of sodium alginate are available from nutritional supplement suppliers.

Calcium Alginate

Calcium alginate is at least as efficient in binding strontium as sodium alginate.[88] Based on animal studies, a modified calcium alginate given to rats over three months, reduced strontium deposition in the skeleton by 70-90%. [89] Another study found a synthesized product containing calcium alginate from red algae, if added to strontium-90 contaminated milk, efficiently fixed the strontium into an insoluble complex.[90] Calcium alginate is available at nutritional supplement suppliers.

---

[87] Memorial Sloan-Kettering Cancer Center, "About Herbs, Fucoidan," www.mskcc.org/mskcc/html/69227.cfm, February 2011

[88] Tanaka, Y, Deirdre Waldron-Edward, Stanley Skoryna, "Studies on Inhibiition of Intestinal Absorption of Radioactive Strontium", Canad. Med. Assn.J, July 27, 1968 Vol 99 p 169 pubmed 5673223

[89] Ivannikov, AT, Altukhova, GA,et al "The Protective Action of Calcium Alginate in Chronic 90 Strontium Uptake into the Body", Radiologia, 1993, Mar Apr 33 (2) 297-311 www.pubmed/8502751

[90] Ivannikov, AT, Altukhova, GA, et al, "Effect of Algisorb on Decontamination of Milk Contaminated with SR 90", Radiats Biol. Radioecol. 1994 July-Oct 34 (4-5) 713-717 www.pubmed/7951905

Chlorella and Spirulina

Spirulina had a radioprotective capability in improving the blood cell production in irradiated mice and dogs. [91]

The same effect was seen in a separate study of mice survival after radiation after administration of Chlorella. [92]Both algae have shown immune enhancing abilities. Chlorella is reported to be a heavy-metal detoxifier. They have shown immune enhancing abilities. They are both available in a variety of forms by nutritional supplement providers.

Flaxseed

Flaxseed increased radiation tolerance of lung tissue in animal studies.[93] Flaxseed has high omega-3 fatty acids and lignans with antioxidant activity. It was also protective of pulmonary fibrosis, inflammation and oxidative lung damage. Flaxseed is available in capsules or as fresh seed for grinding at nutritional supplement providers.

Antioxidants Research

Many antioxidants have anti-mutagenic properties in addition to free radical scavenging according to animal studies.

---

[91]Zhang, HQ, Lin, AP, Sun Y, et al "Chemo and Radioprotective Effects of Polysaccharide of Spirulina Platensis on Hemopoietic System of Mice and Dogs, Acta Pharmacol, Sin, 2001, Dec 22 (12) 1121, pubmed 11749812

[92]Rotkovska, D, Vacek, A, et al "The Radioprotective Effects of Aqueous Freshwater Algae (Chlorella kesslen) in Mice and Rats", Strahlenther Onkol, 1989, Nov 165 (11) 813 pubmed 2688154

[93] Lee, JC, Krochak, R, Blouin A, Kanterakis, S et al,, "Dietary Flaxseed Prevents Radiation Induced Oxidative Lung Damage, Inflammation and Fibrosis in a Mouse Model of Thoracic Radiation Injury", Cancer Biol. Teher. 2009 Jan 8 (1) 47-53 pubmed/18981722

Antioxidants may also protect cells by increasing natural antioxidant GSH (glutathione). Antioxidants ability to reduce cellular damage caused by ionizing radiation has been studied for 50 years.

- In one animal study, antioxidant diet supplementation started 24 hours after exposure reduced radiation lethality by improving bone marrow cell survival. Irradiation at 8 Gy in mice without antioxidants was lethal by 16 days. When the antioxidant was started at 24 hours, 14 of 18 mice were alive and well at 30 days. [94]

- An animal study found that two separate dietary antioxidant formulations had major suppressive effects in the later stages of radiation-induced carcinogenesis. No malignant tumors were observed in irradiated animals maintained on either antioxidant diet. The authors hypothesized the antioxidants prevented the early stage neoplastic growths from progressing and the antioxidants were very effective in preventing the development of tumors induced by radiation changes.[95] One of the antioxidant formulas contained; I-selenomethionine, (SeM), N-acetyl cysteine (NAC), ascorbic acid, coenzyme Q10, a- lipoic acid, and Vitamin E succinate.

---

[94] Brown, SL, Kolozsvary, A, Liu J, et al, "Antioxidant Diet Supplementation Starting 24 Hours After Exposure Reduces Radiation Lethality", Radiat. Res. 2010 Apr 173 (4) 462-3 www.ncbi.nlm.nih.gov/pubmed/20334518

[95] Kennedy, AR, Ware, JH, Carlton, W., et al "Suppression of the later stages of radiation-induced carcinogenesis by antioxidant dietary formulations" Radiat. Res. 2011, Jul; 176 (1) 62-70 pubmed/21520997

Superoxide Dismutase

Superoxide dismutase are enzymes produced by the body as natural antioxidants to protect cells from reactive oxygen damage. White blood cells produce the enzyme to kill bacteria, and some bacteria produce the enzyme for protection. Many antioxidant supplements promote the production of superoxide dismutase.

A recent study found the first six months of tai chi exercise stimulated increased endogenous antioxidant enzymes, including, superoxide dismutase, and reduced oxidative damage. [96]

Superoxide dismutase is available in cream or sprays at nutritional supplement providers. It was found by this author in a high quality facial lotion.

White and Green Tea

White and green tea, contain a higher quantity of catechins compared to black tea and have radioprotective effects. A study in mice revealed antioxidant activity and reduction of inflammatory cytokines and improved recovery of the haematopoietic system.[97] (blood producing system)

Zingerone

Cooked ginger, contains zingerone, which significantly protected lymphocytes from DNA damage by radiation

---

[96] Goon, JA, Aini, AH, Anem, MV, et al, "Effect of Tai Chi Exercise on DNA Damage, Antioxidant Enzymes, and Oxidative Stress in Middle-aged Adults", J. Phys. Act. Health,. 2009, Jan 6 (1), 43-54

[97] Hu, Y, Cao, JJ, Liu, P, Guo, DH et al "Protective Role of Tea Polyphenols in Combination Against Radiation-Induced Haematopoietic and Biochemical Alternations in Mice", Phytother Res 2011 Mar31 doi; 10.1002/ptr 3483 pubmed/21452375

by scavenging free radicals and inhibiting of radiation induced oxidative stress. [98]

We found zingerone is not commonly available, but is an added ingredient of some nutritional formulations.

Anti-inflammatory Drugs

Radiation creates excess production of eicosanoids. (prostaglandins, proacyclin, thromboxanes, and leukotrienes). Eicosanoids are mediators of inflammation that cause vasodilation, vasoconstriction, and chemotaxis, Some researchers suggest glucocorticoids inhibit eicosanoid synthesis and NSAIDs (nonsteroidal anti-inflammatory drugs) prevent prostaglandin/thromboxane synthesis. One researcher suggested they should be studied further for drug treatment of radiation damage to organs.[99]

Aspirin and NSAIDs (ibuprofen, naproxen and others) down-regulate the eicosanoids to some degree, and that is how they reduce fever, clotting, and inflammation. Eicosanoids are produced by the body from Omega 3 and Omega 6 fatty acids. They are controlled in the body by dietary fat and insulin.

Several studies have found NSAIDs may protect normal tissues from radiation and prevent radiation induced side effects. The NSAIDs have great potential to minimize adverse effects of radiotherapy[100]

---

[98] Rao, BN, Archana, PR, Aithal, BK, et al, "Protective Effect of Zingerone A Dietary Compound Against Radiation Induced Genetic Damage and Apoptosis in Human Lymphocytes", Eur. J Pharmacol, 2011, Apr 25:657(1-3) 59-66 pubmed/21335001

[99] Michalowski, AS, "On Radiation Damage to Normal Tissues and Its Treatment II Anti-inflammatory Drugs", Acta Oncol, 1994, 33, (2) 139-57 pubmed/8204269

[100] Lee, TK, Stupans I. "Radioprotection: the NSAID's and Prostaglandins", J. Pharm. Pharmacol. 2002 Nov 54 (11) 1435-45, 12495545

Gastrointestinal Remedies

Antacids, laxatives, emetics, and activated charcoal are gastrointestinal treatments that may be advised by physicians after a radiological emergency. The Chernobyl caregivers encouraged aggressive treatment of gastrointestinal symptoms.

You may wish to keep an emergency supply of these over the counter remedies. Medical authorities may advise the use of some of these agents after they determine what radiation hazard is present.

Other Dietary Supplements Research

A review of radioprotection research included pharmacological management with Vitamin E, Vitamin C, beta carotene, and ribose-cysteine.

Lycopene pretreatment in rats significantly decreased DNA damage from gamma radiation.[101]

L-carnitine may protect against gamma radiation induced cataracts by increasing Superoxide Dismutase activity and scavenging free radicals, based on radiation-induced cataracts in rats.[102]

Ginkgo Biloba was found to have strong antioxidant properties against oxidative stress generated by gamma radiation. [103]

---

[101] Srinivasan M, Sudheer, AR, et al "Lycopene as a Natural Protector Against Gamma Radiation Induced DNA Damage, Lipid Peroxidation and Antioxidant Status in Primary Culture of Isolated Rat Hepatocytes in Vitro", Biochim Biophys Acta 2007, April 1770 (4) 659-65 pubmed/ 17189673

[102] Kocer, I, Taysi, S, et al, "The Effect of L-Carnitine in the Prevention of Ionizing Radiation-Induced Cataracts: A Rat Model", Graefes Arch CLin Exp Ophthalmol. 2007 Apr;245(4):588-94 pubmed 16915402

Aged garlic extract contains antioxidant phytochemicals allixin and selenium that protected DNA against free radical mediated damage and mutations and defended against ionizing radiation.[104]

Pretreatment with curcumin protected hepatocytes (liver cells) against gamma radiation based on rat studies.[105] A study in mice found the curcumin increased mice survival after radiotherapy through antioxidant defenses and decreasing radiation-induced lung fibrosis.[106]

Aloe vera leaf extract protected skin against radiation-induced biochemical alterations in a study of mice. It increased superoxide dismutase and decreased destructive lipid peroxidation in the liver.[107]

Resveratrol is a strong radical scavenger and activated a specific response in irradiated animals who were pre-treated with resveratrol. It is potentially an effective protecting agent, according to the researchers.[108]

[103] Okumus, S Taysi S, Orkmez, M., et al, "Effects of Oral Ginkgo Biloba Supplementation on Radiation Induced Oxidative Injury in the Lens of Rat", Pharmacogn Mag, 2011, April 7 (26) 141-5 www.ncbi.nlm.nih.gov/pubmed/21716624

[104] Borek,. C., "Antioxidant Health Effects of Aged Garlic Extract", J Nutr. 2001, Mar ; 131 (3s) 1010S-5S www.ncbi,.nlm.nih.gov/pubmed/11238807

[105] Srinvasan M., Sudheer, AR, "Effect of Curcumin analog on Gamma Radiation Induced Cellular Changes in Primary Culture of Isolated Rat Hepatocytes In Vitro", Chem. Biol. Interact. 2008, Oct 22 176 (I) 1-8 pubmed 18597748

[106] Lee, JC, Kinniry, PA, Arguiri, E, "Dietary Curcumin Increases Antioxidant Defenses in Lung, Ameliorates Radiation Induced Pulmonary Fibrosis and Improves Survival in Mice", Radiat.Res. 2010, May; 173 (5) 590-601 pubmed/20426658

[107] Goyal, PK, Gehlot, P, "Radioprotective Effects of Aloe Vera Leaf Extract on Swiss Albino Mice Against Whole-Body Gamma Irradiation", J. Environm Pathol Toxicol. Oncol, 2009;28(1);53-61, pubmed 19392655

[108] Ye, K, Ji, CB, Lu, XW, Ni, YH, et al "Resveratrol Attenuates Radiation Damage in Caenorhabdidtis elegans by Preventing Oxidative Stress", J. Radiat. Res, 2010;51(4);473-9pubmed/20679743

Homeopathic Radiation Hormesis?

A 2006 expert panel at the Ultra-Low-Level Radiation Effects Summit in the U.S. proposed a new laboratory to confirm or discard the hormesis treatment sometimes used by homeopathic physicians. The premise of hormesis is that a tiny amount of the damaging agent, induces a protective response from the body. In dose response trials hormesis would produce a non-linear association at low doses.

However, the BEIR VII report continues to support a linear association even at the lowest levels.

A recent study of lung cancer risk in radon exposure found a non-linear association between exposure and the odds of lung cancer. The authors theorized on the striking hormetic dip, and the possible causes, perhaps an activated adaptive biological response. The researcher acknowledged it could also be explained by random variation and needs further investigation.[109]

Some studies found that pre-exposure to radiation causes cells to turn on protective mechanisms.

However, the National Research Council states there is no threshold of exposure of low levels of ionizing radiation that can be demonstrated harmless or beneficial.

Animal studies have not uniformly supported the hormetic response.

---

[109] Thompson, RE, "Epidemiological Evidence for Possible Radiation Hormesis from Radon Exposure; A Case Control Study Conducted in Worcester, MA, " Dose Response, 2010, Dec 14:9 (1) 59-75 pubmed/21431078

Chapter Eight

Fallout Exposure

Is fallout part of the detonation?

Fallout will occur from any size nuclear detonation. It will be larger if the detonation occurred closer to the ground. After a blast, the fallout period will begin within 10 minutes near the detonation. Fallout may start 34 minutes later if the detonation is 10 miles away, based on modeling experiments. Survivors should be wary of "rainout", when the radiation level is increased by precipitation.

What is fallout?

The fallout elements emitting gamma and beta radiation, attach to airborne particles of dust composed of the remains of buildings, concrete and roads in the blast area. The fallout can appear as sand, dirt or dust in the air or on the ground.

In a nuclear power plant accident contamination could be in the steam or ash plume released from the nuclear fuel.

The winds transport the fallout. The upper atmosphere winds may not resemble the ground winds, so you cannot predict the direction of fallout. The winds shift direction and areas previously safe can be affected. Fallout can burn your skin, and it can be inhaled or ingested internally and cause a lifelong health risk.

What is the duration of fallout?

The level of contamination will be highest in areas near the detonation area and downwind. Based on modeling experiments, it will likely peak at about six hours and then diminish gradually over a few days. Eighty percent of the fallout is expected to occur in the first 24 hours.

Survivors should remain sheltered until authorities announce it is safe to evacuate outdoors. According to some sources, the fallout radiation will be reduced by 90% every seven hours.[110] The fallout duration can be extended based on the distance from the event and weather factors. The longer victims avoid being exposed to the outdoor contamination, the better.

Based on modeling, in a few days, the fallout zone can diminish from 10 miles to 2 miles. Depending on the size of the discharge or blast, it could take 30 days for the fallout radiation to decay to safe levels.

How is fallout measured?

A personal radiation detector is useful during the fallout period. A radiation dose of 0.01 R/h is a "radiation caution zone" and requires protective equipment. A respirator and suit is required for emergency responders. Within hours of the blast, the 0.01 R/h level can extend several hundred miles.

---

[110]U.S. Dept. of Health and Human Services, Public Health Emergency, ":State and Local Planners Playbook for Medical Response to a Nuclear Detonation", www.phe.gov/Preparedness/planning/playbooks/stateandlocal/nuclear/pages/backgrou nd.aspx

10 R/h is considered a "hot zone" and limits access by emergency workers. Based on modeling of 10Kt nuclear detonations in various locations, at a distance of 1.6 miles it took 17 hours to reduce the radiation level to less than 10R/h. At a distance of 3.4 miles, it took 13 hours to reach 10R/h. At 5.9 miles it took 8 hours and 30 minutes to reduce the radiation to less than 10R/h. [111]

The radiation level is expected to increase to a peak at 4-6 hours and then to drop. Time will eventually reduce the volume of radioactive particles in the air. There is a "7-10 rule." A seven-fold increase in time is expected to cause a tenfold decrease in the radiation rate.

Federal authorities have the ability to monitor radiation levels remotely. When the outside contamination is safe, emergency services will arrive. If you do not observe emergency workers, assume it is unsafe and stay in your adequate shelter. If you observe emergency workers in protective suits, there is still contamination present. In this situation, if you must go outside, you can make your own protective garment with garbage bags and duct tape, including your shoes. Seal all seams with the tape. Authorities may decide to move victims in spite of fallout, if your shelter is inadequate.

Does the government still maintain fallout shelters?

No. The program ended in 1965. In 1961, the Federal government established a Community Fallout Shelter Program. The local communities selected the buildings that provided a protection factor of 40 (40 times better protection than outdoor exposure). The space was

---

[111]Buddemeier, BR, Dillon, MB, "Planning Factors for the Aftermath of Nuclear Terrorism", Aug 2009, www.remm.nlm.gov/ResponsePlanning_LLNL_TR-410067.pdf section 313.

sufficient for 50 people and the shelters were stocked with two weeks of supplies. Local communities maintained the shelters and the supplies until the 1970s. You may be able to find what buildings were designated as fallout shelters from historical records in your library. In the 1960s, public buildings were generally unlocked. In a current emergency event, finding an unlocked building may be an obstacle for shelter. The Planning Guidance Report in 2010 recommended that communities with inadequate structures designate radiation shelters.

How is self-decontamination performed?

The shelters can be contaminated with radiation fallout by the dust particles on clothing, shoes and hair of evacuees.

Dusting off, or removing and bagging the contaminated clothes outside of the shelter will reduce contamination inside. Contaminated clothing normally removed over the head should be cut off to avoid contact with the eyes, nose, and mouth. Once inside an adequate shelter and no longer exposed to fallout, you can begin decontamination of the skin and hair. Some shelters may require decontamination before you enter. Radiation emergency planners do not recommend that shelter be delayed for decontamination.

As an added precaution, victims should wash their hair, blow their nose and wash their eyelids and mouth.

If no water is available, use wet wipes to clean especially around the mouth, nose, ears and eyes and skin. Your emergency kit should include wet wipes.

Bare skin should not be cleaned with brushes, abrasives or harsh chemicals. The skin is an important barrier to avoid internal contamination. The mildest cleaning solution should be used initially. If cleansers are not effective or causing irritation, use hand lotion and then try cleaning again.

Open wounds could allow radioactive particles to penetrate internally. Wounds should be flushed with clean water to remove fallout particles and covered. Skin disinfectant solutions that contain iodine are an effective wound cleaner.

What self reliance emergency training is available?

The American Civil Defense Association (TACDA) has a free tutorial at their website, www.tacda.org. We encourage readers to review the lessons and to support the non-profit organization.

The association is a non profit organization to empower Americans with a comprehensive understanding of reasonable preparedness strategies in the event of nuclear, biological, chemical or other disasters. They were not contacted regarding this book, but we support their mission as stated on their website in educating the public in emergency preparation.

How can the reader assist family and community?

The reader will have heightened awareness of the exposure risks and shelter options in their area and should share this information as needed. When the reader purchases emergency kit supplies, they should consider extra inventory for those in the same shelter. Sharing radiation detector readings after an event could also benefit the community. Providing assurance to others in shelter about the intentional delay in rescue operations and the priority to schoolchildren will be beneficial.

We also recommend that readers carry a bible in their radiation emergency kit.

Chapter Nine

Shelter

Why shelter in place?

The survival rate is increased by three simple measures, increasing time, distance and shielding. When the radiation measures 100mR/hr, then after 3 hours it would be 300 mR exposure. Where the exposure at one foot is 500 mR/hr, the exposure at two feet is 125mR/hr. Any shielding will protect against alpha radiation and any solid shielding from beta radiation. Gamma radiation will be reduced by lead and concrete shielding especially in right-angle configuration. Deep soil cover will also provide shielding.

If you are caught outdoors, an underground shelter may be the best option.

Fallout will begin from eight to thirty minutes, depending on the distance from the blast. You must be in your adequate shelter before the fallout cloud arrives. It is unlikely that you can reach your ideal or pre-stocked shelter in this time-frame. You must identify the best option immediately.

What is the best available shelter?

You may have to find shelter by memory, so it is a good idea to evaluate potential shelters in advance. During your commute to an urban area you should monitor substantial buildings along the route that may provide emergency shelter. Look for armories, banks and hospitals and seek underground locations. Look for tunnels, subways, or parking garages on your commute to work.

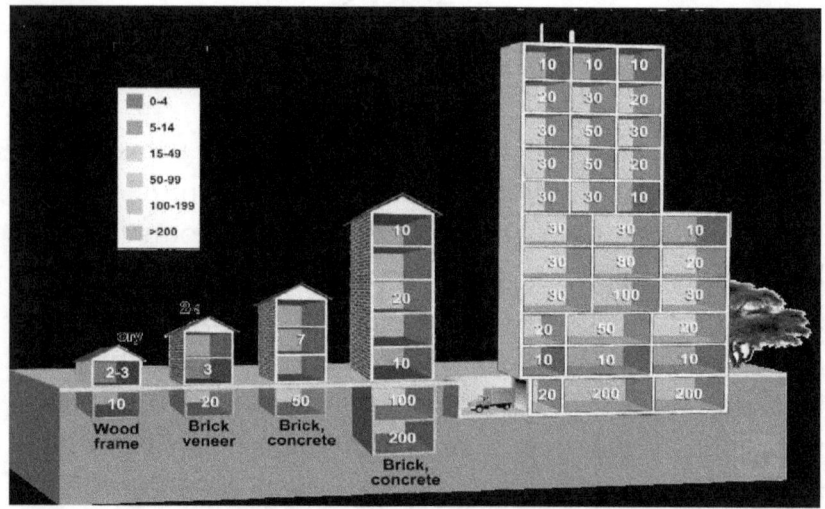

Buildings as shielding- numbers represent a dose reduction factor. A dose reduction factor of 10 indicates that a person in that area would receive 1/10th the dose of a person in the open.[112]

You should evaluate your office building, home and any other frequented location based on the relative safety of this graphic. If your building is inadequate, look at other structures you could gain access. The safest buildings may have been historical fallout shelters, buildings with thick brick or stone walls and buildings with basement access or interior rooms without windows.

A deep basement offers good protection. You need to have protection overhead and from windows and entrances. The middle floors of multi-story buildings and

---

[112]"Planning Guidance for Response to a Nuclear Detonation 2nd Edition", http://org/hsc/documents/planning guidance for response to a nuclear detonation-2nd Edition_Final.pdf p 75, www.remm.nlm.gov/Planning GuidanceNuclearDetonation.pdf

interior rooms without windows are the best if you cannot reach the basement.

What does protection factor mean?

The dose reduction factor is described as "PF" for Protection Factor. The PF measures the protection at three feet above the surface. A good shelter has a protection factor of 100. Adequate PF is 10 and inadequate is 3. [113]

- A vehicle has a protection factor of 1.7-2. Vehicles are not adequate shelter.

- At a PF of 3, such as a one-story frame house with no basement, two hours exposure could be a lethal dose. It is still better than an unsheltered location where a lethal dose could be received in 15 minutes.

- A PF of 10, such as a brick building without a basement can protect you significantly from a lethal dose for at least 3 ½ hours.

- A PF of 100, such as the inner rooms of a large office building or a deep basement, will protect you from lethal radiation.

- Three feet underground has a protection factor of 5000.

  Even if you are in a single-story frame house without a basement, stay in the center at ground

---

[113]Buddemeier, BR, M. B Dillon, "Planning Factors for the Aftermath of Nuclear Terrorism", Aug 2009 Lawrence Livermore National Laboratories www.remm.nlm.gov/ResponsePlanning-LLNL_TR_410067.pdf

level until the fallout has passed. Moving from bad shelters during fallout is dangerous.

What are the risks of leaving a shelter?

Based on a modeling of a 10 Kt detonation, if a person leaves a shelter after 15 minutes, they will have a 250 rem exposure. This is nearly a lethal dose. If they wait 24 hours, they may limit the exposure to less than 2 rem. [114] It is best to stay in a good shelter with a PF of 100, for one to four days based on modeling experiments.

If your radiation detector measures 10 rads/hour or more, or you are in an inadequate shelter, you need to increase the level of shielding. Sacks of rice, books, containers of water, and dirt near the entrances or windows will enhance shielding. These shields will also enhance basement shelters, if placed on the floor above the shelter.

If a person is in an inadequate shelter with a PF of 10, they may reduce exposure by enhancing the protection by creating a barrier inside of an inner room with books, tables or any available thick surface. They may move to a better shelter after a few hours. They should know that a better shelter is in close proximity before attempting to leave.

If a person is in a poor shelter, such as a vehicle or frame house with no basement, they should try to improve their shielding immediately with materials inside of the shelter and seek better shelter in close proximity after the fallout cloud passes. Based on modeling and distance from the detonation this might occur within one hour.

---

[114]Buddemeier, BR, M.B. Dillon, "Planning Factors for the Aftermath of Nuclear Terrorism", Aug 2009, Lawrence Livermore National Laboratories www.remm.nlm.gov/Responseplanning_LLNL TR-410067.pdf section 2.4.3.

Can you build a fallout shelter?

The shelter would need to be constructed before a radiation event.

It is reasonable to build a fallout shelter in a corner of your basement. It provides a high level of protection, is easy to build and will be usable for years. A personal fallout shelter will be useful for those who are at home when the event occurs. You will probably not have time to get home before lethal fallout occurs after a detonation. It is best to shelter-in-place in an adequate shelter.

A basement fallout shelter in your home could be located in a corner, away from windows, and near the highest outside ground level. You need to add two concrete block walls in the basement and add radiation protection to the ceiling. The general floor layout could be as shown below:

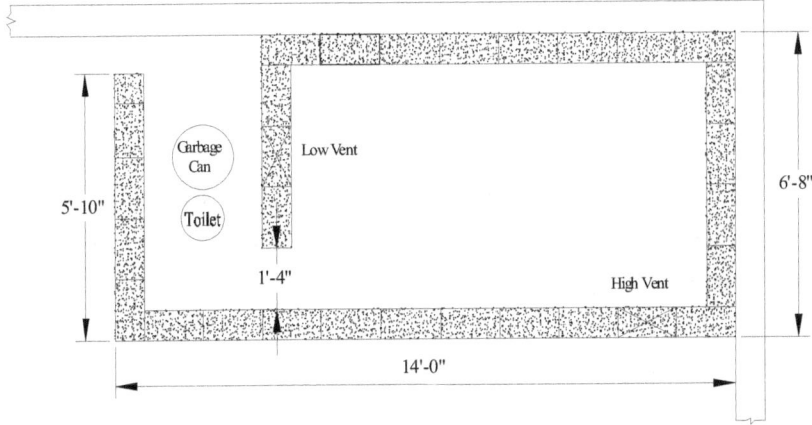

The fallout shelter can be made longer or shorter depending on your needs. This layout insures that there is no direct path for gamma radiation through the walls.

The high vent includes a fan to pump filtered air into the room.

A vestibule is useful for decontaminating before entering the enclosed area or as a bathroom in a long term shelter, to keep this area separate from the living area. Include a sealed bucket, kitty litter, several garbage bag liners and lid for this purpose.

The ceiling can be low, since you will be sitting or sleeping in the room.

Secure strong beams across the top of the shelter walls to allow you to add plywood to make a level surface on the top. Add sideboards around the perimeter of the top so that you can put sand or dirt on top of the shelter to add additional protection from gamma radiation.

The fallout shelter needs to be narrow to provide structural support in case your house collapses. The fallout shelter could include bunks, tables and chairs. Attach hooks and use sheets or blankets to make hammocks if more beds are needed.

If you do not have a basement, decide if you can use a ground level room or if you need to dig a shelter. You can add radiation protection for the windows and other openings using dirt or thick materials. You can store sand bags to wall a small area within the room. Perhaps, placing a table as a ceiling, over the walls of sand bags. Materials can be placed on the table top to protect against descending gamma radiation. Improvise with the materials that you have.

If you have a basement, but no pre-built shelter, you can plan to improve the protection. Consider what materials you have to build the inside walls. Bookcases filled with books will provide some protection against gamma radiation. A pool table, turned on its side, could provide protection. You should have lumber and tools ready that could be used to quickly modify a shelter.

What are emergency power needs?

After any nuclear detonation, you should plan for loss of power due to EMP (electromagnetic pulse). More information will be presented later on the risk of EMP. A battery can provide power for LED lights, an emergency radio, and the ventilation fan. The ventilation fan in your shelter needs to run 24 hours a day and is the largest drain on the battery.

The table on the next page shows the watt-hour per day requirement for basic power appliances at about 576 watt-hours. The power estimate for the fan assumes two of the do-it-your-self HEPA filters described in the Ventilation Chapter. The two small 12 volt computer fans should be adequate airflow for five persons. Adjust this if you have higher or lower ventilation requirements.

Device	Hours per day	Device Power watts	Watt-hours Per day
LED lights	16	5	80
Radio	8	2	16
Fan	24	20	480
Total			576

Since we may need to stay in the shelter for up to three days, a total of 576 x 3 = 1728 watt-hours of energy are needed. At 12 volts, this corresponds to 1728 / 12 = 144 amp-hours. Two deep cell marine batteries should be able to provide power for your fallout shelter if they are fully charged. Use a battery maintainer to keep the batteries safely charged in preparation.

What emergency supplies should be maintained?

Your basement shelter can be used as storage for your emergency supplies. It is a good idea to store two weeks of food and water for the maximum number of occupants. The two-week supply of food should not require heating or elaborate preparation. Some authorities even recommend a year's supply of food.

In addition, include tools like shovels, crowbars, and saws so you can dig yourself out, if the house collapses above you. Be sure to include an emergency radio, CB radio, fire extinguisher and flashlights.

A radiation detector, even a build it yourself, Kearny Fallout Meter, is essential to determine radiation levels. You can use this tool to determine which area of your shelter has the least radiation. For example; the lower three feet may have a reduced radiation level compared to the upper three feet. The fallout meter will help in determining how soon you can leave your shelter, and for how long of periods.

Will the water be safe to drink?

City water should be safe as long as the city water supply tank is enclosed. Immediately after an emergency, fill any container with water for storage. The water supply may eventually be depleted or contaminated. Covered water sources should be free of radiation.

If there is no source of safe water, do not obtain water from a surface source, like a creek or lake. The better alternative is to dig down to the water table. Three feet below the surface there may be a perched water table above an impermeable layer of rock or clay. It should be filtered and disinfected before using. Clay soil will filter many contaminants to some degree including radiation particles. Clay soil has densely packed small particles. It smears when pressed when wet. Clay forms below the topsoil. At three feet below the ground, dense clay should be free of radiation.

If you are unable to dig into the ground water, you can remove some of the radiation by filtering potentially contaminated water through the fresh clay. For drinking you should further purify the water against bacteria, by boiling or adding 8 drops of household bleach per gallon of water. You can check the level of chlorine with the chlorine test strips in your emergency supplies before drinking to insure it is no more than 4 ppm. 15 gallons of water per person should be stored for a two week shelter

Bottled water packaged before the radiation event is the best source. Hopefully, the emergency operations will be able to provide bottled water for shelters.

What foods are safe to eat?

Plants growing outdoors will become contaminated after a nuclear accident/attack. For long-term needs, grow vegetables inside, away from radiation. Use non radioactive water to clean vegetables. Fruits should have the outer skin cut away.

The safest foods to eat are pre-packaged and produced before the nuclear accident/attack. Keep a supply of food and water for emergencies.

What guidance is available for building a fallout shelter?

The American Civil Defense Association (TACDA) has a book available, Technical Directives for the Construction of Private Air Raid Shelters, at www.tacda.org.

Several designs for fallout shelters are reportedly available from FEMA.(Federal Emergency Management Agency) . They have a publication "Design Guidance for Shelters and Safe Rooms in Buildings." This FEMA Publication 453 was published in 2006. They also have a manual, "Safe Rooms and Shelters, Protecting People Against Terrorist Attacks." Other FEMA publications are TR 29 and TR 87 that may be useful. Readers should research all available designs to prepare a shelter.

Readers are advised to contact a professional contractor to determine what shelter construction is suitable for your location and home. The FEMA publications that we could find on the internet focused on administrative rather than practical applications.

The shelter and detection lessons taught in the text, Nuclear War Survival Skills by Cresson H. Kearny are still relevant.

The website www.homelandcivildefense.org has practical information and more recommended plans. Historical military and civil defense sites also have information on fallout shelters.

An underground fallout shelter provides the best protection against radiation from a nuclear attack or major nuclear accident. Former federal agencies and historical civil defense organizations have designs available on-line. Most shelters are designed to have a protection factor of at least 40, which is the minimum standard of protection for public shelters throughout the United States. The American Civil Defense Association recommends a PF of 1000. Adding 24 inches of dirt surrounding the shelters will increase the protection factor significantly.

The following are fallout shelter designs from historical sites. The websites credited with the diagrams usually provide details of the construction. Readers are encouraged to research these sites for more detailed information.

The illustrations reveal the age of the designs.

Underground Fallout Shelter

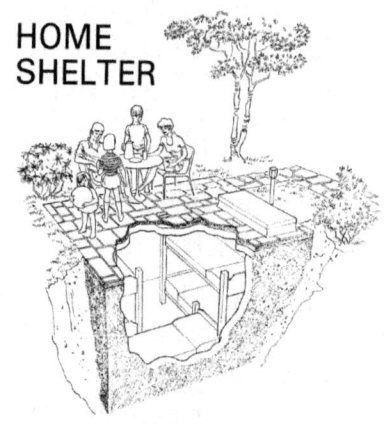

HOME
SHELTER

images.military.com/ContentFiles/cw_shelterh121.pdf

The above design is capable of housing six adults. It can be built of poured concrete, precast concrete slabs, or a combination. If it is built as detailed, with the top near ground level, the roof can be used as an outdoor patio, as shown in the picture above. The shelter is accessible by a hatch-door and wood stairway. [115]

Fresh air is provided by a hand-operated centrifugal blower and ventilating pipes that extend aboveground level. If your water table is high, the fallout shelter should be built aboveground level to prevent flooding. Be sure to add earth around all sides of the shelter to protect against gamma radiation.

---

[115] http://images.military.com/ContentFiles/cw_shelterh121.pdf

Above ground Fallout Shelter

www.cddc.vt.edu/host/atomic/pdf/sheltr02.pdf

This family shelter is intended for persons that prefer an aboveground shelter. The shelter is designed to meet the standard of protection against fallout radiation. It can also be constructed to provide significant protection from the effect of hurricanes, tornadoes, and earthquakes. This design provides limited protection from the blast and fire effects of a nuclear blast. It has sufficient space for six adults[116].

The shelter can be built of two rows of concrete blocks. One 12" and one 8," filled with sand and grout, or of poured reinforced concrete. This structure has been designed for areas where frost does not penetrate the ground more than 20 inches. If 20 inches is not sufficient depth for footings, one or two additional courses of concrete may be used to lower the footings. Average soil

---

[116] http://www.cddc.vt.edu/host/atomic/pdf/sheltr02.pdf

bearing pressure is 1500 lb per sq. ft. Most soils can be assumed to support this pressure without special testing or investigation. The wood frame over the reinforced concrete ceiling probably would be blown off by extremely high winds such as caused by a blast wave or tornado. However, the wood frame is intended primarily for appearance.

Home Fallout Shelter Concrete Block Shelter basement location

images.military.com/ContentFiles/cw_sheltrh12c.pdf.

The above design is a compact shelter that can be installed in a basement corner using common lumber and concrete blocks with mortar joints for permanent construction. This shelter has about 50 square feet area, and will provide shelter for five persons.[117] Its purpose is to provide adequate protection for a minimum cost in an existing basement. The materials required to build this

---

[117] http://images.military.com/ContentFiles/cw_sheltrh12c.pdf

shelter are obtainable through local concrete block plants and/or home supply stores. Natural ventilation is provided by the entrance-way and the air vents in the shelter wall. The chapter on ventilation will provide details on filtering the air.

Home Fallout Shelter Modified Ceiling Shelter-basement location

www.cddc.vt.edu/host/atomic/pdf/sheltr03.pdf

The above fallout shelter should be used only in low risk areas. A low-risk area is one which is not expected to be subjected to the blast effects of a nuclear weapon. This shelter can be permanently installed in the basement of your home and will not interfere with its utility[118].

In basements whose walls are mostly below grade on all four sides, adequate shelter from fallout radiation is

[118] http://www.cddc.vt.edu/host/atomic/pdf/sheltr03.pdf

provided by adding additional protection on the ceiling. Plywood is screwed to the bottoms of the joists to support masonry shielding. A beam and jack post is used to support the additional weight. In a one-story house, approximately one-quarter of the area of the basement ceiling should be filled with concrete blocks or bricks in order to obtain the most protection out of this design. A building contractor will provide assistance in designing the sufficient support structure.

Home Fallout Shelter Ceiling Shelter-basement plan B

www.civildefensemuseum.com/southrad/manuals/shelter plans/h-12-b.pdf.

This design should only be constructed in low risk areas. A low-risk area is one which is not expected to be subjected to the blast effects of a nuclear weapon. New

2x12 joists, notched to the depth existing 2x10s, are installed alongside these joists in order to carry the extra weight of the shielding material[119]. This eliminates the need for a beam and jack post to support the ceiling as used in the previous design.

Shelter with tilt-up storage unit

www.civildefensemuseum.com/southrad/manuals/shelter plans/h-12-e.pdf.

The principal feature of this shelter is a roof composed of tilt-up storage units, the top of which is hinged to the wall. The units can be used as book cases, pantry shelves, or for miscellaneous storage. In an emergency, the storage units can be tilted up so that they rest on a stacked masonry wall built from materials stored nearby the units.

---

[119] http://www.civildefensemuseum.com/southrad/manuals/shelterplans/h-12-b.pdf

In basements where the outside ground level is above the top of the tilted-up units, adequate shelter from fallout radiation is provided by filling the units with brick or solid concrete block 8" thick. This shelter will house six people[120].

Lean-to Basement Fallout Shelter

www.civildefensemuseum.com/southrad/manuals/shelter plans/h-12-f.pdf

This shelter is designed to provide protection from the effects of radioactive fallout in the below the grade basement of an existing house. Its advantages are low cost, simplicity of construction, general availability of materials, and the fact that it may be easily disassembled.

---

[120] http://www.civildefensemuseum.com/southrad/manuals/shelterplans/h-12-e.pdf

This shelter design will provide 54 square feet of area and approximately 216 cubic feet of space. It will house three persons. The shelter length can be increased by increments of 3-foot panels. Height increase will be limited by basement height and handling of the panels[121].

Is filtered ventilation needed?

A chapter will follow later about ventilation options. A 1962 report found that air filters were not necessary in small family shelters. Inadequate ventilation is a greater risk than unfiltered air intake. Ventilation designs are included later. The Kearny Air Pump design is available on many websites and in the book <u>Nuclear War Survival Skills,</u> by Cresson H. Kearny.

What are recommendations for those who self evacuate after the radiation levels are safe?

Even if you have created a shelter for your family, it is likely that some family members will be sheltered elsewhere. When you choose to leave a shelter, move perpendicular (right angle) from the direction of the fallout or steam plume in the air. You should have an emergency contingency plan for meeting with your family at a few potential locations that have adequate protection. You may have pre-positioned supplies at these locations. If CB radios have not been affected by an EMP, you can communicate with your family. You must have pre-arranged times and frequency of transmissions for radio contact.

---

[121] http://www.civildefensemuseum.com/southrad/manuals/shelterplans/h-12-f.pdf

Chapter Ten

Decontamination of the Environment

What can be done for radiation contaminated soil?

After the contamination of miles of good farmland, Chernobyl researchers attempted methods to remove the radiation from the ground through mechanical and chemical methods. Some of these methods included;

- Application of potassium fertilizer to reduce cesium uptake.

- Adding lime to increase calcium levels and reduce strontium 90 uptake.

- Planting grains in place of leafy vegetables and pasture because grains accumulate lower levels of radionuclide.

- Sugar beet or rapeseed crops that are processed and decontaminated after production.

- Deep plowing to dilute soil contamination

- Planting flax, cotton, rape seed for bio fuel, or ornamental plants where the products are not consumed [122]

- Testing of crops, and if contaminated, plowed under.

- In Belarus and the Russian Federation, prussian blue supplements were added to the diet of dairy and meat animals causing a radiation reduction factor of up to 10. In the Ukraine a clay mineral binder was used, and not as effective.

Fertilizers containing potassium can decrease the ratio of cesium to potassium and reduce the plant root uptake of cesium. Liming of the soil, discing and fertilizing were effective on some soils based on soil pH, nutrition, and soil type. The contamination reduction factor was 2-4 for poor soil and 3-6 for organic soil.

The Chernobyl decontamination counter-measures were cost effective when applied soon after the accident. They found removal of the soil layer, conducted in previous accidents, was too expensive. Other radiation remediation methods were:

- Contaminated grazing cattle were moved and slaughtered.

- Rejected contaminated milk was sent for processing for condensed milk and cheese or butter.

Was the contamination successfully remediated?

[122] Richards, JI, "One Decade After Chernobyl: The FAO Response", Keynote Presentation, IAEA Chernobyl Update, April 1996

A 2008 U.N Scientific Committee reported that during the first year after the Chernobyl accident, there was a fast decline in radioactivity in the soil, but limited reduction over the next four to six years.[123]

The U.N. report found cesium concentrations in these food products in descending order of contamination; potatoes, grain, milk and meat. As of 2008, fresh fish in Scandinavia contained radioactive cesium still higher than action levels.[124]

In the highly contaminated areas, 80% of the grain and milk exceeded standards at the end of 1986. By 1991, less than 0.1% of the grain had cesium levels above 370 Bq/kg.[125] The U.N report found contamination from food was more significant than from drinking water and fish consumption.

How were environmental areas decontaminated?

Gross decontamination steps that removed 20-95% of the contamination from roads, porches, etc. are described as;

- vacuuming,

- fire hosing,

---

[123] United Nations, Scientific Committee on Effects of Atomic Radiation, UNSCEAR Chernobyl's Legacy Health Environment and Socio-economic Impacts", Chernobyl Forum 2003-2005 www.iaea.org/Publications/Booklets/Chernobyl/Chernobyl.pdf

[124] United Nations, Scientific Committee on Effects of Atomic Radiation, UNSCEAR Chernobyl's Legacy Health Environment and Socio-economic Impacts", Chernobyl Forum 2003-2005 www.iaea.org/Publications/Booklets/Chernobyl/Chernobyl.pdf

[125] United Nations, Scientific Committee on Effects of Atomic Radiation, UNSCEAR Chernobyl's Legacy Health Environment and Socio-economic Impacts", Chernobyl Forum 2003-2005 www.iaea.org/Publications/Booklets/Chernobyl/Chernobyl.pdf

- washing with detergents,

- steam cleaning,

- sand blasting,

- vegetation, soil removal and

- road resurfacing.[126]

Chernobyl decontamination actions were removal of trees and gardens, washing asphalt, concrete, and building walls and roofs. The dose rate was reduced by 1.5 to 15 times.[127]

The surface waters had high radiation concentrations after the accident, but they fell rapidly. In most settlements in 2008, the air was normal, except for over undisturbed soil.

The highest cesium was found in forested areas. Contaminated wood from forests burned for alternative heating fuel was a cause of radioactive exposure.

Forest produce of mushrooms, berries, game, reindeer and fish from shallow lakes had the highest cesium concentrations.[128]

How was zeolite used for decontamination?

[126] National Security Staff Interagency Policy Coordination Subcommittee for Preparedness and Response to Radiological and Nuclear Threats, "Planning Guidance for Response to a Nuclear Detonation", June 2010 2nd edition "The Radiation Treatment, Transport and Triage concept"p 63
www.remm.nlm.gov/PlanningGuidanceNuclearDetonation.pdf

[127] United Nations, Scientific Committee on Effects of Atomic Radiation, UNSCEAR Chernobyl's Legacy Health Environment and Socio-economic Impacts", Chernobyl Forum 2003-2005 www.iaea.org/Publications/Booklets/Chernobyl/Chernobyl.pdf p,39

[128] United Nations, Scientific Committee on Effects of Atomic Radiation, UNSCEAR Chernobyl's Legacy Health Environment and Socio-economic Impacts", Chernobyl Forum 2003-2005 www.iaea.org/Publications/Booklets/Chernobyl/Chernobyl.pdf p 53

There are forty varieties of naturally occurring zeolites. Zeolites are microporous minerals found in volcanic soils. They are used in a variety of industries, including water purification, laundry detergent production, soil treatment, and pet odor control.

On hard surfaces, zeolite has been used in nuclear accident clean-up efforts. Zeolite has a history of use for radiation and metal detoxification;

- Synthetic zeolite was used to take up strontium and cesium from the waters near Three Mile Island nuclear plant accident.[129]

- Clinoptilolite zeolite, added to soil contaminated with strontium 90, reduced the strontium uptake by plants. Clinoptilolite inhibited the uptake of cesium in Bikini Atoll soil, contaminated by nuclear tests. [130] (Clinoptilolite is a naturally occuring zeolite)

- Zeolite treated pasture land near Chernobyl reduced the absorption of cesium by sheep [131]

Other sources have reported using zeolite in animal food supplements. Zeolite apparently exchanges cesium and strontium in the gastrointestinal tract and the radiation is then excreted by normal processes, minimizing assimilation into organs.

---

[129] Mumpton, Frederick, A "La Roca Magica: Uses of Natural Zeolites in Agriculture and Industry", www.pnas.org/content/96/7/3463.full?sid=b070df77-7159-4ab5-8e4e-794f849284f0

[130] Mumpton, Frederick, A "La Roca Magica: Uses of Natural Zeolites in Agriculture and Industry", www.pnas.org/content/96/7/3463.full?sid=b070df77-7159-4ab5-8e4e-794f849284f0

[131] National Academy of Sciences, Geology, Mineralogy and Human Welfare", "Nuclear Waste and Fallout" 1998 Proceedings of the National Academy of Sciences, www.pnas.org/content/96/7/3463.full?sid+3bcf8a83-3968-4c2a-a832-f4f6fdd75de2

Zeolites are selective for strontium, cesium, cobalt, calcium and chromium.

How can shelter areas be decontaminated?

The best option is to hire trained professionals to perform decontamination of living and working areas.

Remember that you are not eliminating the radiation, only moving it to a different, hopefully safer, location.

If decontamination cannot wait, as in a crowded shelter, the following may be useful. Decontamination involves damp mopping. Start with the least contamination and work towards the most contaminated area. Discard the used damp mop cloth often, and replace it. Wear protective suits over clothing and shoes. Wear eye protection and respiratory masks. If there are no other resources available, you can create a protective suit with plastic garbage bags and duct tape. Make certain to cover your shoes also.

Ordinary detergents and clean water can be used, according to our reading. Mop and wipe surfaces slowly to avoid disturbing the dust. Change damp rags frequently. Discard the used cleaning cloths and treat it as radioactive. Test for radiation after cleaning with your radiation detector.

Ordinary vacuuming is not generally recommended because of distribution of the radioactive particles into the air and contamination of the machine. However, some sources say a vacuum with a HEPA filter could be employed for dry contamination.

Chapter Eleven

Measuring Radiation

How many radiation measurement units are there?

There are several units used to measure radiation: the Roentgen (R), the Roentgen equivalent man (rem), the rad (radiation absorbed dose), the gray (Gy), and the Sievert (Sv). An "m" before any of these unit symbols means one-thousandth. For example: 1,000 mSV = 1Sv. A μ before any of these symbols means one-millionth, 1,000,000 μSv = 1 Sv.

Each of the measurement units above defines an amount of radiation. For example: a mammogram exposes a person to 2 mSv of radiation. The amount of radiation an individual receives from background radiation is 2mSv/year. Therefore, one mammogram exposes a person to the same radiation level received during a whole year from normal background radiation.

Why are there so many measurements?

The reasons there are so many units for radiation measurement are historical, industry-specific, international, or an intent to measure the harmful effect of radiation on a person. Older texts use R, rem, or rad for measuring radiation level. Newer texts will use Gy and Sv as defined by the Système International.

The Sievert (Sv) and the Roentgen equivalent man (rem) radiation measurement units attempt to measure the harmful effect of radiation on living tissue. These are called "dose equivalent units". For example, if you are exposed to 1 Gv of gamma radiation, your dose equivalent exposure is 1Sv. However, if you have been exposed to 1 Gv of alpha particle radiation, your equivalent exposure is 20 Sv. Alpha particles are more massive and damaging to tissue than gamma rays.

- Rad - not adjusted for additional harmful effect.

- R - not adjusted for additional harmful effect.

- Gy - not adjusted for additional harmful effect.

The dose equivalent measurement for Sv and the rem are computed as follows:

Dose Equivalent = Absorbed Dose x Quality Factor

- rem - adjusted for additional harmful effect.

- Sv - adjusted for additional harmful effect.

How do you determine the relative risk of different types of radiation?

Quality factor adjusts the value to account for more destructive particles. The quality factor for some common radiation agents follows:

- X-ray or gamma rays   1

- Beta particles   1

- Neutrons (unknown energy)   10

- Protons (unknown energy)   10

- Alpha particles   20

- Particles of unknown energy   20

Alpha particles are 20 times worse than beta particles or gamma rays. Plutonium 238 emits alpha particles and produces a high quality factor in the formula.

How can you compare one set of measurements with another?

For most other situations, you can assume a quality factor of approximately 1. With this assumption, the following can be calculated:

1 R = 1 rem = 1 rad = 0.01 Sv = 0.01 Gy

1 Gy = 1 Sv = 100 rem = 100 R = 100 rad

Convert between units using the following table:

- 1 rem = 0.01 Sv = 10 mSv

- 1 mrem = 0.01 mSv = 10 μSv

- 1 Sv = 100 rem

- 1 mSv = 100 mrem = 0.1 rem

- 1 μSv = 0.1 mrem

All of the above information assumes that you are exposed to radiation from a source outside your body. However, you can inhale radioactive particles into the lungs, or they can be ingested in food or water. Radioactive particles can lodge in the lungs and remain for a long time. As long as they remain and continue to decay, the exposure continues.

How do you measure the quantity of radioactive material present?

The Becquerel (Bq) and the Curie (Ci) are measurements used for the quantity of radioactive material.

- Bq- One Bq is defined as the <u>activity</u> of a quantity of radioactive material in which one <u>nucleus</u> decays per <u>second</u>.

- Ci- The <u>curie</u> (Ci) is an older unit of radioactivity equal to the activity of 1 gram of radium-226. The typical human body contains roughly 0.1 μCi (millionth of a Curie) of naturally occurring potassium-40. A typical radiotherapy machine may have 1000 Ci of a radioactive material. This can cause serious health effects in minutes if you are directly exposed. The following table shows how to convert between Ci and Bq.

$1 \text{ Ci} = 3.7 \times 10^{10} \text{ Bq}$

$1 \text{ Ci} = 37 \text{ GBq}$

$1 \text{ } \mu\text{Ci} = 37{,}000 \text{ Bq}$

$1 \text{ Bq} = 2.70 \times 10^{-11} \text{ Ci}$

$1 \text{ Bq} = 2.70 \times 10^{-5} \text{ } \mu\text{Ci}$

1 GBq = 0.0270 Ci

Notice that 1 Curie is much larger than 1 Becquerel. This requires the use of large prefixes for conversion. The small number subscript means to multiply the number, in the first example below, (10), six times for positive numbers, or to divide that number six times for negative numbers. (10 X 10 X 10 X 10 X 10 X 10)

1 MBq (megabecquerel) $= 10^6$ Bq

(10 X 10 X 10 X 10 X 10 X 10)

1 GBq (gigabecquerel) $= 10^9$ Bq

(10 X 10 X 10)

1 TBq (terabecquerel) $= 10^{12}$ Bq

How is contaminated land measured?

The level of radiation on a land area can be expressed as Bq/m^2 or Ci/Km^2. For example, the evacuation zone after the Chernobyl nuclear accident had CS^{137} deposits with 555,000 Bq/m^2 ($15Ci/km^2$) of radioactivity. The level of radioactivity in the land can be used to estimate the external radiation dose that a person living in that area would receive. It is estimated that a person living in the Chernobyl evacuation zone would receive 7mSv of radiation from exposure to Cs 137. Convert land area units using the following equivalents.

1 Bq/m^2 $= 2.7$ x 10^{-5} (0.000027) Ci/km^2

1 Ci/km^2 $= 37000$ Bq/m^2

What are recommended radiation limits?

There is a wide range of radiation dosage limits based on occupational exposures. The National Council on Radiation Protection and Measurements states;

- 10R/hr is the fallout zone delimiter. [132] No work inside of the perimeter of that contamination level is recommended unless it is sufficiently justified.

- 50 rem as a possible limit but mission specific rescues at 10.[133]

The EPA 1992 Public Action Guide Manual states

- 25 rem is the limit of exposure for emergency workers,

- 5 rem is the normal occupational dose limit unless missions are on a voluntary basis. [134]

[132] National Security Staff Interagency Policy Coordination Subcommittee for Preparedness and Response to Radiological and Nuclear Threats, "Planning Guidance for Response to a Nuclear Detonation", June 2010 2nd edition "The Radiation Treatment, Transport and Triage concept" p53
www.remm.nlm.gov/PlanningGuidanceNuclearDetonation.pdf

[133] National Security Staff Interagency Policy Coordination Subcommittee for Preparedness and Response to Radiological and Nuclear Threats, "Planning Guidance for Response to a Nuclear Detonation", June 2010 2nd edition "The Radiation Treatment, Transport and Triage concept" p 52,59
www.remm.nlm.gov/PlanningGuidanceNuclearDetonation.pdf

[134] National Security Staff Interagency Policy Coordination Subcommittee for Preparedness and Response to Radiological and Nuclear Threats, "Planning Guidance for Response to a Nuclear Detonation", June 2010 2nd edition "The Radiation Treatment, Transport and Triage concept" p 55

How is radiation monitored in emergencies?

Radiation detectors can be used to measure the current radiation level. Not all detectors measure all types of radiation or all levels of radiation.

A dosimeter records the accumulated radiation. These detectors are usually cards or stickers that are small and inexpensive.

A radiation meter or Geiger counter can record the current radiation level. Survey meters can record the radiation level at various locations and times.

For radiation survival, you want to know both the current radiation level and the total accumulated exposure level.

The current radiation level warns you of areas to avoid. The total accumulated level of radiation warns you of the health risk exposure.

The following are some of the radiation detectors that we found on an internet search. We recommend that you purchase a detector of your choice, after thorough research. You may also build your own radiation monitor based on the design by Cresson H. Kearny. The source of those designs are described later in this book

The detectors on the market that follow, are provided to show you a sampling of the range of options available. You should research these products and others thoroughly before purchase. Some dosimeters only measure gamma radiation. You will want to select a meter that measures a wider range of fallout radiation threats while you are in your shelter.

SIRAD™

SIRAD™ (**S**elf-indicating **I**nstant **R**adiation **A**lert **D**osimeter) is a badge radiation detector for monitoring high doses (1 rad to 1000 rad) of ionizing radiation. This type of radiation detector is low cost and does not need batteries. There is no alarm if you are suddenly exposed to radiation. The card must be inspected visually to determine radiation exposure. Also, the card records the total accumulated radiation, not the current exposure level. For example; if you look at your card once a week and notice a high reading compared to a week ago, you won't know when or where that radiation exposure occurred. You could have had a single exposure sometime during that week. Or, you could have had a low level exposure that accumulated over the week. A SIRAD™ will not monitor a radiation dose below the lowest level indicated on the card.

The main advantage of the SIRAD™ is that it is inexpensive and small enough that you can carry it with you at all times. If you are a first responder or work in an area where you could be exposed to high levels of radiation, you should definitely carry a SIRAD™ (or similar dosimeter) card.

If you live near a nuclear power plant you should carry a dosimeter card. If you work in a major city or a target, you should carry a dosimeter card. Remember to check the card one or twice daily. If you have any reason to suspect possible exposure, check the card every few minutes.

RadSticker™

The RADSticker™ peel and stick is a postage stamp-sized dosimeter that you can stick on the back of your driver's license or anything you keep close, and ready for a radiation emergency. The RadSticker™ can measure exposures from 25 rad to 1000 rad.

RADTriage™

This is a credit card-sized 3-3/8" x 2-1/8" dosimeter. The shelf life is about one year. It measures radiation from 500 to 10,000 mSv. After one year, the card will develop a color equivalent to about 10 mSv. No battery or calibration is required.

Nukalert™

The Nukalert can be used as a key chain and carried around in your pocket or purse. The unit makes a faint ticking sound to let you know it is working. This is a radiation detector that measures the current rate of exposure, not the accumulated exposure that a dosimeter records. It measures a range of 100mR/hr to 50R/hr. This is a radiation detector that is small enough that you can carry it with you all the time.

palmRADII Nuclear Radiation Meter

This is a small, hand-held radiation meter that detects a wide range of exposure levels 100 urem/h to 300 rem/h. The palmRADII includes data logging to record radiation levels at various locations. The unit runs on 4 standard "AAA" batteries. The battery life is 20 hours from fully charged batteries.

This type of radiation meter is most useful to first responders that need to record where and when radiation levels are high. For personal use, the requirement to recharge the batteries every 20 hours will limit its utility.

Radiation Alert Inspector

This is a small one pound hand-held radiation detector that will record 1uR/hr to 100 mR/hr. This unit includes an audio alert function, as well as standard Geiger counter ticker functionality. It is powered by a single 9-volt alkaline battery that is good for six months.

Gamma-Scout® Alert Radiation Detector

This is a compact six-ounce radiation detector that includes a large display and 10 button control panel. It has a built-in clock/calendar with data memory and logging capability. It includes both ticker and alarm mode to warn of radiation. It will measure both weak and high radiation levels: 0.01 uSv/h to 1000 uSv/h. This unit includes an internal battery that will reportedly last ten years.

Radiation Alert® Digilert-100®

The Digilert-100® measures alpha, beta, gamma, and X-radiation. Use it to check personal radiation exposure, monitor an area or perimeter, detect radiation leaks and contamination, or identify changes in background radiation. An audible, pulsating alert sounds when the radiation reaches a user-selectable alert level. The recording range is from 0.001 mR/hr to 100 mR/hr.

Rad Scanner 500

The Rad Scanner 500 is a small six digit Geiger counter with solid state electronics. This is a pocket sized unit. It will run two months on a single 9 volt battery. The Rad Scanner includes both visible and audible indicators. The alarming range can be set from 0 to 999mR/hr.

Terra MKS-05

This is a Ukrainian-built radiation detector that has the ability to switch between current exposure rate and total accumulated dosage. The unit is powered by two AAA batteries and has a three month operating life.

Radex 1706

This is a Russian-made unit with minimal features. A backlit LCD panel displays updated readings every 10 seconds. Ticker and threshold alarm functions are included. The unit is powered by a pair of AAA batteries, which will provide three weeks of continuous operation.

Are civil defense Geiger counters still an option?

Thousands of Radiation Survey Meters and Geiger counters were manufactured starting in the 1950s. These units need to be maintained and calibrated periodically to maintain accuracy. Unless you know that a surplus unit is working correctly and has been calibrated, it is best to avoid these units.

New radiation detectors are sold at reasonable prices, are smaller, have a longer battery life, and are generally better than the surplus units available. In addition to purchasing a radiation meter, you can build your own Kearny Fallout Meter.

How do you build a Kearny Fallout Meter?

The Kearny Fallout Meter is an accurate meter that you can build yourself out of commonly available items. It was developed by Cresson H. Kearny at Oak Ridge National Laboratory. The design is simple, so there is nothing to calibrate and no electronics to break. This meter does not need batteries and will not be affected by an EMP.

The meter depends on measuring the collapse of two aluminum foil leafs that have been charged with electricity. The higher the radiation level, the faster the charged leafs will collapse. The design for the Kearny Fallout Meter is available on the Internet at: www.oism.org, www.ki4u.com or http://www.cddc.vt.edu/host/atomic/pdf/kfm_inst.pdf and many other sites. These plans and more practical ideas are in <u>Nuclear War Survival Skills</u> by Cresson H. Kearny. There are also kits and completed units available for sale on the Internet.

Because of the size of this book and the restrictions that the templates cannot be changed, we are unable to provide the plans in this book. We feel confident that you will be able to locate the plans on the internet or in his book.

We built a detector based on the plans. We made small modifications when building our meter. We have no way of testing the effect of our modifications on the radiation detection. We found the lid was difficult to make as described in Kearny's instructions. We found the lid from an Activia parfait yogurt container fits tight and is sturdier than the sheet plastic recommended. We found that 2-lb fishing line works best for the thread holding aluminum foil and for the stop threads that keep the foil from opening too wide.

We used flower preserving silica gel for the drying agent, rather than heated drywall as the plan recommends. We used Dri- Splendor brand which includes blue crystals that turns pink when it is no longer effective. You can "re-dry" the drying crystals by heating them for 30 to 45 minutes at 250 degrees.

Can plants be used to measure radiation?

If you have no monitor, Tradescentia pallida, is a plant that is reportedly sensitive to radiation and can be used to detect radiation. It is used in Brazil and other tropical climates. Studies conducted at Kyoto University in Japan and at Brookhaven National Laboratory found that the normally blue stamen hairs mutate by turning pink when exposed to radiation. [135]

---

[135]Dos Santos Leal TC, Crispim, VR, Frota, M et al, "Use of a Bioindicator System in the Study of the Mutagenetical Effects in the Neighborhoods of Deposits of Radioactive Waste" Appl. Radiat. Isot. 2008 April 66 (4) 535-8 pubmed 18164207

Some non-scientific sources refer to the plant as Spiderwort. The plant exhibits mutational effects at low doses of ionizing radiation. Since 1974, T. pallida have been planted around nuclear power plants in Japan. About 12 million stamen hairs are observed every year in Japan for radiation monitoring. Significant increases in mutation frequencies were correlated with operation periods of nuclear facilities and wind direction. [136]

Other plants have also been tested in the laboratory to determine the relative degree of contamination in the area. One study at the Savannah River Site found that the degree of cesium-137 contamination could be determined in pine needles. Plutonium-238 and 239 were elevated in bur-reed plant roots. Bladderwort measured extremely high levels of cesium and plutonium.[137]

[136] Ichikawa, S, "In Situ Monitoring with Tradescantia Around Nuclear Power Plants", Environ. Health Perspect. 1981, Jan 37, 145-64

[137]Caldwell, EJ, Duff MC, Ferguson CE, et al, "Plants as Bio-monitors for cs-137, Pu-238, 240, and K-40 at the Savannah River Site", J. Environ Monit. 2011, May 13 (5) 1410 Mar 16 pubmed 21412545

Chapter Twelve

The Government Role

How did the EPA respond to the recent nuclear accident?

In review, the Fukushima nuclear power plant experienced a power failure and back up power depletion on March 12, 2011. Radioactive steam was released to cool the radioactive fuel. Between March 21 and 22, radioactive cesium 134, cesium 137 and iodine 131 was present in EPA samples. Some stations measured uranium in the air. These were reported in the EPA RadNet Air Filter results, issued on 4/4/2011.

The EPA stated there was no cause for alarm. The levels were below any level of public health concern. Radiation neutralizers or decontamination were not recommended.

However, the BEIR VII report, states there is no safe level of radiation.

How many organizations control radiation standards?

The World Health Organization, International Atomic Energy Agency, (IAEA), the International Commission on Radiological Protection (ICRP), United Nations Scientific Committee on the Effects of Atomic Radiation (UNSCEAR), and the National Council on Radiation

Protection and Measurements (NCRP), are some of the international agencies that produce radiation reports and recommendations.

Is the response to radiation accidents affected by public psychology or socio-economic considerations?

The agencies and organizations are subject to political and economic pressures from industry and from the public. Some of their reports have been contradictory. Early Chernobyl reports were criticized for underestimating the damage. Some agencies presented the affected population as having an unreasonable fear of radiation. They were reluctant to link any medical conditions to the accident. The possible dismissal of diseases and the description of the population as "radiophobic" is a lesson in the potential bias in government sponsored evaluations.

The World Health Organization (WHO) reported in 1992, "it is important to reduce public anxiety following a nuclear emergency, since psychological pressures exert a strong influence on public behavior at the time of an accident and after." They recommended a consultation on the psychosocial dimensions of contingency planning. [138]

The WHO bulletin also reported that "perception of radiation risk by the public is often greater than the actual risk and this may lead to misrepresentation of information."[139]

The International Commission on Radiological Protection encouraged planning for radiological attacks. The

---

[138] World Health Organization, Bulletin of the World Health Organization, 70, (3) 391-392

[139] World Health Organization, Bulletin of the World Health Organization, 70, (3) 391-392

summary adds, in order to prevent "over-reaction", response measures prepared in advance should reflect the real expected gravity of the various possible scenarios.[140] Does this imply intentional planning for a minimal event? The new U.S. Planning Guidance is based on a 10Kt detonation, (a low to moderate size device).

The International Commission on Radiological Protection reported that after a radiological emergency, the immediate countermeasures are primarily caring for people with traumatic injuries and controlling access. They describe later actions, as respiratory protection, personal decontamination, sheltering, iodine prophylaxis if needed[141], and temporary evacuation. These plans are very similar to the U.S. Planning Document.

The Belgian radiation plan may provide some insight into the perspective of government emergency plans that are not available for the public to view in many countries.

According to the Belgian plan, a group at the Coordination and Crisis Centre in Brussels will make political decisions based on both "radiological and socio-economic considerations." "Protective measures may be proposed by the radiological experts but will probably be modified in light of socio-economic considerations such as the social and or economic disruption that might arise" [142]from deploying proposed measures.

---

[140] Valentin, J., "Protecting People Against Radiation Exposure in the Event of a Radiological Attack A Report of the International Commission on Radiological Protection, Ann ICRP, 2005,35(101-110, iii www.ncbi.nlm.nih.gov/pubmed/16164984

[141] Valentin, J., "Protecting People Against Radiation Exposure in the Event of a Radiological Attack" A Report of the International Commission on Radiological Protection, Ann ICRP, 2005,35(101-110, iii www.ncbi.nlm.nih.gov/pubmed/16164984

[142] Van Bladel, L, Vandecasteele, Cl, "Organization of the Nuclear Emergency Plan, Verh K Acad Geneeskd Belg 2005;67(5-6) 337 www.ncbi.nlm.nih.gov/pubmed/16408829

A lecturer on "Informing the Public about Radiation—the Messenger and the Message," stated there are many obstacles to radiation protection communication. He stated there is considerable disagreement within the profession about the content of the message to the public. He added, the objective is to ensure that everyone recognizes that radiation protection opens the door to the benefits of the applications of radiation in medicine and industry. [143] He rightly encourages plain and consistent language in the science and education of the public.

One writer, representing the nuclear industry, reminded those in the radiation protection field that they have been less than successful in one of their most important tasks, effective communication. The writer was concerned the public was afraid of radiation, even at low doses.

An UNSCEAR report of 2008 stated the Chernobyl victims were exposed to radiation levels comparable to or a few times higher than annual levels of natural background and future exposures continue to slowly diminish as the radionuclides decay. Apart from thyroid cancer and indications of increased leukemia and cataracts, the report said there was no clearly demonstrated increase in the incidence of solid cancers, leukemia due to radiation in the exposed population. [144]

What is the federal response to radiation emergencies?

---

[143] Lakey, J, "Informing the Public About Radiation", Health Phys. 1998 Oct; 75 (4) 367 www.ncbi.nlm.nih.gov/pubmed/9753359

[144] UNSCEAR 2008 Report Vol II, p176 D.pdf www.unscear.org/docs/reports/2008/11-80076-Report_2008_Annex

The U.S. Department of Defense is utilized for nuclear events under the National Response Plan. Details of that plan are not available to the public. The Planning Guidance reveals much about the federal response, the categorization of victims, and the intentions of search and rescue efforts.

The US EPA has a Radiological Emergency Response Team (RERT) with a mobile response laboratory that can reach any location in the U.S. reportedly in 2-4 days to assist in radiation accidents. They also issue Protective Action Guides, (PAG's) to assist local emergency responders in making decisions after a nuclear attack or radiation accident. They have a graphic that assists in determining the response. It is not too difficult to understand. Victims must shelter and decontaminate. The dilemma is that the public must take the action before the federal authorities arrive.

The US Department of Energy produces guidelines and provides medical consultation through the Radiation Emergency Assistance Center. They produced a technical report for physicians on treatment recommendations. It includes a recommendation for psychological staffing because of the irrational public reaction to radiation.

The following is a protective action guide. [145]

Exposure Pathways and Protective Actions

These are examples of exposure routes and various protective actions. The phases are not set timeframes and protective actions may overlap more than one phase.

POTENTIAL EXPOSURE PATHWAYS	INCIDENT PHASES	PROTECTIVE ACTIONS
1. External radiation from facility	EARLY	1. Sheltering, evacuation, control of access
2. External radiation from plume		2. Sheltering, evacuation, control of access
3. Inhalation of activity in plume		3. Sheltering, administration of stable iodine, evacuation, control of access
4. Contamination of skin and clothes		4. Sheltering, evacuation, decontamination of persons
5. External radiation from ground deposition of activity	INTERMEDIATE	5. Evacuation, relocation, decontamination of land and property
6. Ingestion of contaminated food, water	LATE	6. Food and water controls
7. Inhalation of re-suspended activity		7. Relocation, decontamination of land and property

Why were essential safety protocols released only to emergency personnel and not to the public?

In June of 2010, a federal "Planning Guidance for Response to a Nuclear Detonation" was released to local and State emergency planners. The document relies on these authorities to inform the public as to preparation, sheltering and evacuation. Although it is intended for nuclear attack, many of the recommendations would also apply to a nuclear accident.

The Planning document has valuable information on the range and duration of nuclear contamination and the relative protection of shelters. The planning document confirms that much of the survivability is based on the

---

[145]US EPA, Radiological Emergency Response Team, "Protective Action Guides", www.epa.gov/rert/pags.html

public knowing instantly what action to take when they see the initial "flash" of light.

This action by the public, requires advance education and preparation. However, ordinary Americans are not the intended audience for the planning guidance document.

We have not heard or seen any practical information relayed to the public since the document was released. The information has high value to the average American before the event, but is wasted if delivered hours after an attack or nuclear accident.

Radiation survival is unique because shelter must be found immediately, wherever the victim is at the time of detonation. A stocked shelter will be useful if you happen to be at home.

Sadly, Americans were better educated on nuclear safety in 1962. At that time, Community Fallout Shelters were posted and stocked with medical supplies, food and water.

The Federal government knows you have a better chance of surviving with advance information, but are reluctant to issue the advice. We do not understand why the federal government does not provide this information directly to the public.

Should you rely on the federal resources for radiation emergencies?

No. After reviewing the federal plan, Americans will realize that in the initial stages, their survival will depend on their own preparation and knowledge. Victims need to stock their own radiation decontamination, neutralization supplies to use when advised by their physicians, build or purchase their own radiation monitor and study the best shelter options.

A well-stocked radiation emergency kit could contain a radiation meter, neutralizing agents, decontamination wipes, water, food, tools, materials for a protective suit, and respiratory mask.

What are priorities in the planning guidance document?

The Planning Guidance prioritizes emergency responders to school locations. The children are to remain in the adequate shelter until the radiation levels are safe for evacuation. Parents are expected to remain sheltered until it is safe to evacuate. This information needs to be shared prior to an event to prevent parents from endangering their own lives to rescue family members.

Planners suggest that family members independently seek shelter and reconnect during the later evacuation stage when the radiation dose is near normal. We agree with the recommendation, but it is not useful if it is not shared with the public.

Rescuer safety is a priority issue, and we support the intent. If the radiation dose is over 5 rem, emergency responders must be fully informed of the risks of exposures and are allowed to work on a voluntary basis and with personal protection equipment.[146]

The Federal Planning Guidance encourages coordinators to balance the risk to responders with the benefits of lives

---

[146] National Security Staff Interagency Policy Coordination Subcommittee for Preparedness and Response to Radiological and Nuclear Threats, "Planning Guidance for Response to a Nuclear Detonation", June 2010 2nd editionwww.remm.nlm.gov/PlanningGuidanceNuclearDetonation.pdf -

saved. This is a reasonable decision, but the victims should know if, when or where they may be rescued. This information should be shared with the public.

What are other recommendations to emergency responders in a radiation emergency?

- Emergency units may be advised to shelter themselves until the radiation dose has fallen.

(Author's note: The radioactive fallout may last from 12 hours to three days. Victims will need water and food at their shelter location. In a nuclear plant release, the radiation may be continuously emitted for a longer time.)

- Encourage victims to seek shelter in safe locations.

(Authors note: Victims need to know the relative safety of a shelter **before** the rescuers arrive)

- Focus search and rescue in non-radioactively contaminated areas.

(Authors note: Non-radioactive areas would be far outside the detonation or nuclear accident)

- Avoid the severe damage zones where the victims are "expectant."

(Authors note: "expectant" means expected to die of radiation, perhaps four to six weeks later.)

- Register victim's identification, screen for radiation, and possibly order decontamination in a separate location before shelter admission, when resources arrive.

(Author's note: documentation protocols may take precedence over shelter and neutralization and will delay shelter safety)

- Federal resources are expected to arrive in one to three days after the detonation.

(Author's note: The maximum fallout is in the first 12 hours. Victims should have advance knowledge and a personal supply of radiation neutralizers. Physicians would recommend action before authorities arrive)

- Close to the blast, in the severe zone, survivable victims are low priority. Responders are advised to avoid this severe zone for at least 12 hours.

(Author's note: Victims will be on their own during this critical time to avoid fallout, and decontaminate. These victims should be informed in advance that assistance may not be available)

- Emergency responders are advised to care for targeted victims in targeted areas. These targets are, severe injuries in the light damage zone and survivable victims in moderate damage zones.

(Author's note: Rescue operations are limited and prioritized by target zones. Victims should know in advance if rescue assistance is expected.)

- Responders need to protect people from lethal doses.

(Author's note: in close proximity, the emergency responders will not be available to unsheltered survivors until radiation levels are already potentially lethal.)

- Electronic media and news will inform victims about shielding and evacuation at the time of the event.

(Author's note: it is highly likely after a detonation that an EMP will destroy all electronic communications for miles. The same EMP will likely disable emergency and escape vehicles and power equipment.)

- Victims should "shelter in place".

(Author's note: We agree, but fallout will occur in less than 10 minutes, near the detonation, long before emergency responders are able to direct victims to shelter)

- Victims must decontaminate to protect the shelter from contamination if possible.

(Author's note: Standing orders for decontamination protocols issued by administrators not on the scene could result in delays for medical evaluation and treatment)

- The urgency of the rescue and the number of survivable victims will be balanced against the rescuer safety of waiting for further radioactive decay.

(Author's note: We support the protection of rescue workers. However, the public should know the criteria for possible rescue and why they may not arrive).

- The emergency responders will have the radiation dose levels, and will avoid locations based on those readings.

(Author's note: The victims' and their families, probably in different locations, must rely on timely and accurate information by emergency responders, unless they have their own radiation meter. We suggest individuals have their own personal radiation detectors.)

- Public shelter operations will screen victims with detectors, order decontamination, enroll victims in long term monitoring programs, and possibly issue identifying wrist bands.

(Author's note: Once inside a public shelter, victims mobility, decontamination and neutralization options will be limited to those offered by the staff and approved by the government.)

- Emergency responders are to reassure the public and "speak with authority". The shelters will be staffed by some government employees who have inservice training in radiation safety.

(Author's note: Readers of this book may be better informed about radiation survival than the assigned disaster staff.)

- The government acknowledges that there will be self-evacuees. The government plans to direct them to a monitoring station for tracking.

(Author's note: the federal government plans to have an evacuee tracking database coordinated by the Agency for Toxic Substances and Disease Registry. They will be asking for addresses, phone numbers, contact information and victim's location at the time of the detonation.)

What happens when the federal government resources arrive in three days?

The Planning Guidance includes a graphic of Radiation Treatment, Transport and Triage that will capture self-evacuators and direct them to Evacuation Centers for transport to outside facilities and expert centers.[147]

Community Reception Centers are planned by the CDC to be opened 6-48 hours after an event. They will have a contamination control zone for initial sorting of victims, registration, first aid, radiation dose assessment, contamination screening, washing and discharge.

The entire Community Reception Center will be located outside the hot zone and staffed by government officials. Washing facilities and wrist bands are described in the decontamination planning guidance. If there is adequate

---

[147] National Security Staff Interagency Policy Coordination Subcommittee for Preparedness and Response to Radiological and Nuclear Threats, "Planning Guidance for Response to a Nuclear Detonation", June 2010 2nd edition "The Radiation Treatment, Transport and Triage concept"www.remm.nlm.gov/PlanningGuidanceNuclearDetonation.pdf

staffing, they plan to have more restrictive radiological screening. [148]

The centers will be staffed by volunteer radiation professionals and government workers with in-service radiation emergency training.

The Department of Health and Human Services will assist with establishing a registry of potentially exposed individuals, performing dose reconstruction and conducting long-term monitoring of the population for potential health effects.

The National Stockpile of Emergency Medicines for emergencies includes; Prussian blue, Ca-DTPA, Zn-DTPA, Ca gluconate, Na alginate, D-penicillamine, Ammonium chloride, Potassium Iodide, and Sodium Bicarbonate. The Department of Homeland Security Planning Guidance for Radiation Incidents will determine what prophylactic or decorporation agents will be provided, including potassium iodide.

In past radiation accidents the neutralizers have not been efficiently distributed.

---

[148] National Security Staff Interagency Policy Coordination Subcommittee for Preparedness and Response to Radiological and Nuclear Threats, "Planning Guidance for Response to a Nuclear Detonation", June 2010 2nd edition "The Radiation Treatment, Transport and Triage concept" illustration p
114www.www.remm.nlm.gov/PlanningGuidanceNuclearDetonation.pdf-p

Chapter Thirteen

The Damage Zones

Who can expect rescue?

Based on the planning document, evacuation, medical care, search and rescue services and staffing are based on the damage zone of the victims location after a 10 Kt nuclear attack. Identifying these zones is useful to a survivor to know if they may receive assistance.

The following are the Planning Guidance descriptions of the damage zones. Based on the descriptions, victims can determine their zone by visual observation of the damage.

Light Damage Zones have broken windows and mostly non life threatening injuries and is the largest affected area. The wind pressure associated with the blast is about 0.5 psi to 3 psi. This zone extends farther geographically but has limited structural damage. Residual fallout radiation is the greatest risk. Based on atmospheric winds, the fallout area could increase for a long time before decaying.

Moderate Damage Zones has rubble, downed utility lines, collapsed buildings, blown out interiors, overturned cars and the most survivable victims. Smoke and dust will affect visibility for at least an hour. The wind pressure may be 3 psi to 8 psi. The wind speed is 150 mph at 5 psi. This zone could extend one mile from a 10 KT weapon detonation location. Unsheltered victims may be affected by the thermal pulse of heat within two miles of the blast. Flying debris will cause most injuries. Unsheltered victims would receive a lethal dose of radiation, 400 rads. Rescuers may target this area when the radiation levels have diminished.

Severe Damage Zones have high radiation levels and destroyed infrastructure. These areas are close to the blast, a 10KT weapon would create a half mile zone of severe damage. The wind pressure could be 8 psi or greater. The wind speed is 300 mph at 10 psi. The thermal pulse of heat that follows the initial blast will cause mortality. Emergency responders will not enter this zone for at least 12 hours. Survivors are possible in the most protective shelters.

Evacuations will not begin until it is safe based on radiation levels. The shelter evacuations will direct evacuees to move at right angles to the fallout plume. The evacuations will be prioritized based on impending hazards, fires, and collapsed structures, medical needs, food and water availability. Those who have poor-quality shelters (of PF 2 or below) and intense radiation may be moved to a better shelter (PF10) by emergency responders.

The Treatment Intervention Window based on a 10 Kt improvised explosive device identifies;

- "Effective medical therapy" is expected at a dose rate more than 2 Gy, but less than 6 Gy.

- "Potentially effective medical intervention" expected in the radiation dose range of 6-18 Gy.

- Over 18 Gy, palliative therapies such as narcotics, fluids, blood products and antibiotics may be administered.

Chapter Fourteen

———

Survival Shelter Ventilation

During the acute fallout period should you reduce exposure to contaminated outdoor air?

During the hours following a radiation emergency, the fallout radiation will be most severe. During at least the first day or when directed by authorities, you will want to avoid exposure to outside air. High quality caulking and weather stripping are best for doors and windows. In an emergency, duct tape can be used to seal any remaining leaks. You should store many rolls of duct tape in your emergency supplies for this purpose.

In your home fallout shelter, you can prepare it in advance. Use plastic to cover the doors and windows as an added barrier. To save time during an emergency, pre-measure and cut the plastic sheeting for each opening.

Make sure the damper for the fireplace is tight. Use duct tape to seal the damper during a survival situation.

Pipes leading to a basement or crawl space can be a major source of air leaks. Seal all these pipe openings with caulk or urethane foam. Electrical outlets are another common source of air leaks. Buy foam inserts to seal these. Ceiling electrical fixtures can be sealed.

Air vents should be sealed to make an airtight house during the initial fallout period. Some vents are part of

HVAC ventilation system and will not draw outdoor air if the HVAC system is shut down. Others, such as bathroom and stove vents, can draw in contaminated outside air. Cover these vents during an emergency to prevent air infiltration.

Urethane foam can be a key component in sealing air leaks. The foam is self-adhesive, resists pressurization from ventilation equipment, and is virtually impermeable to air.

Single-component urethane is dispensed as a bead for gap and crack filling, while two-component urethane is dispensed as a spray or a stream to fill larger holes and voids. A "gap" is up to 3" wide, and a "hole" is anything larger. A "crack" is less than 0.25" wide, and may be sealed with one-component foam or caulk.

The basement has a great many areas for potential air leaks. Use a generous amount of urethane foam to seal this area.

Isn't it necessary to have ventilation to breath?

Fresh air will eventually be needed to provide sufficient oxygen and expel carbon dioxide. Even if you live in an apartment, you can add filtered air intakes to protect you against a poisoned atmosphere. Authorities report that you need 11 sq. ft of floor space per occupant to maintain healthy carbon dioxide levels.

Asphyxiation (respiratory death) is from too much carbon dioxide, not necessarily too little oxygen. Typical indoor air contains 0.06% carbon dioxide. You may start to feel drowsy at a 1% concentration of carbon dioxide. At 2%, you may notice a feeling of heaviness in the chest and require deeper and more frequent breaths. The breathing rate doubles at 3% carbon dioxide, and is four times normal at 5%. At levels above 5%, carbon dioxide is toxic. A 10% carbon dioxide level is lethal in 30 minutes.

Mild carbon dioxide poisoning symptoms are:

- Feeling drowsy
- Frequent and deep breathing
- Muscle twitching
- Flushed skin
- High blood pressure

High levels of carbon dioxide poisoning symptoms are:

- Headache
- Lethargy
- Irregular heartbeat
- Panic
- Convulsions
- Unconsciousness
- Eventually death

You can estimate the time it would take to reach 2% carbon dioxide by taking the volume of air times 0.02 and divide by the production of carbon dioxide for all the occupants. The production of carbon dioxide is about 50 cubic feet per day per person. Applying this method for a 2000 square-foot home (16,000 cubic feet of air) with four occupants calculates to 1.6 days or 38 hours. The following table gives an estimate of the time it would take to reach 2% carbon dioxide levels for different size houses and number of occupants.

	2 persons	4 persons	8 persons	12 persons
500 ft^2 house	8 hours	4 hours	2 hours	1 hours
1000 ft^2 house	19 hours	9 hours	4 hours	2 hours
1500 ft^2 house	28 hours	14 hours	6 hours	3 hours
2000 ft^2 house	38 hours	19 hours	9 hours	4 hours

Time to reach 2% Carbon Dioxide in completely airtight house

Based on the increase in carbon dioxide, ventilation will be needed to keep the carbon dioxide at a safe level. Fresh air should be brought in from the outside faster than the rate the occupants are exhaling carbon dioxide. An air intake rate of 50 / 0.02 = 2500 cubic feet per day is needed to equal the production rate of carbon dioxide.

This corresponds to about 2 cfm (cubic feet per minute) of fresh air. With a safety factor, you should have an input airflow of 5 cfm per person to prevent the buildup of carbon dioxide. This is less than the 15 cfm per person minimum recommended during normal operation, but should be adequate. However, in hot weather you will need more airflow to reduce heat buildup in the room.

A properly designed energy-efficient house will meet air flow requirements during normal operation. However, if power is out for an extended time, or you intentionally shut it off to reduce contamination, you must rely on natural ventilation. We stated previously in this text, filtered air is not a requirement for home shelters.

How can you safely increase outdoor air when necessary?

When the fallout is no longer obvious as dust, but you are unsure if it is safe, one option is to place furnace filters in opened windows to filter the air before it enters the house. It will not filter all the contaminants, but it is better than carbon dioxide poisoning when you need fresh air.

Buy filters to fit your windows ahead of time. A fan could be used to move air near the window opening if there is not sufficient natural airflow.

In case you were considering balancing the oxygen/carbon dioxide with plants, it is not feasible. It would take 150 mature trees to exchange the carbon dioxide of one person.

How do you prevent carbon monoxide?

Carbon monoxide is generated by open flames. Carbon monoxide is a colorless, odorless, tasteless gas that is very difficult for people to detect. Symptoms of mild acute poisoning include light-headedness, confusion, headaches, vertigo, and flu-like symptoms. Use a carbon monoxide monitor to test the level of this dangerous gas.

Any open flame in an airtight house is dangerous. The carbon monoxide level can build up to dangerous levels.

Do not even use candles or other open flames when the house ventilation is shut down due to fallout.

Carbon monoxide buildup is a potential problem in any room that is made airtight. This is particularly true if you use candles or have a catalytic heater or a wood fire. You should have a carbon monoxide detector that reads CO level in parts per millions. These are relatively inexpensive at $30 and could save your life. The following table gives the risk level for CO poisoning:

0-9 ppm (parts per million) CO: no health risk; normal CO levels in the air.

10-29 ppm CO: problems over long-term exposure; chronic CO problems such as headaches, nausea

30-35 ppm CO: flu-like symptoms begin to develop, especially among the young and the elderly

36-99 ppm CO: flu-like symptoms; nausea, headaches, fatigue or drowsiness, vomiting

100 ppm + CO: severe symptoms; confusion, intense headaches; ultimately brain damage, coma, and/or death

Can you use a whole-house HEPA filter to remove radioactive particles?

HEPA filters are very efficient at removing particles from the air. The simplest way to prepare a room for protection, including a biological attack is to use a high-quality HEPA air purifier in the room. Select a HEPA unit that has high enough capacity to filter the air in the room every hour. However, a significant amount of air pressure is needed to force air through a HEPA filter. A typical HVAC system will not be equipped to provide enough air pressure to use a HEPA filter.

In order to use a HEPA filter you need to install a separate HEPA filtration system. A system like this will pull some of the air out of the regular air flow through your furnace, and then boost the air and pass it through a HEPA filter. The extremely filtered air is then returned back into the normal air flow.

The HEPA system can be installed on the existing duct work. A typical whole-house HEPA system is designed to clean and filter the volume of air in an average 2200 sq. ft. home in about an hour.

Aren't there ventilation systems intended for shelters?

There are commercial fans for bomb shelters that will reportedly protect against all known airborne toxins: nuclear, biological, or chemical. Overpressure relief valves are included in the design to maintain positive air pressure in the fallout shelter room.

However, these units are costly, and the fans require 4.3 amps at 12 volts. A survival room may need to be able to operate independent from the power supply for several months. Providing this much power adds considerable size and cost to the alternative off-grid power system.

How can you find the plans for the Kearny Air Pump?

Cresson H. Kearny, who developed plans for the Kearny Fallout Meter, also created plans for a Kearny Air Pump. The plans are available at the Oregon Institute of Science and Medicine website, www.oism.org and in the text of Nuclear War Survival Skills by Cresson H. Kearny.

Are there portable shelters with ventilation for disasters?

Another option, if you do not want to build a fallout shelter is to buy or build an overpressure tent. These are tents designed to be used indoors to protect from nuclear, biological, or chemical attack by maintaining a positive inside air pressure. Air is filtered before it enters the tent. A tent and associated blower can be stored away until needed. This is a great option if you rent or live in an apartment and cannot modify the building. This is also portable, so if you plan to sell soon, you can take your protected shelter with you.

Some companies such as Hardened Shelters LLC at hardenedshelters.com make complete tent systems that store away in a large 154 lb crate when not needed. Their LSS-80 model has space for six persons for an extended period of time. This unit is designed to remove biological warfare agents, including Anthrax, remove all blister and nerve agents, and remove radioactive iodide gas potentially released in a nuclear power plant accident. However, a thin walled tent will not protect you from gamma radiation and provide little protection from beta particles.

The air filtration system used in the LSS-80 is reportedly very effective at removing contamination. A pre-filter, HEPA filter, and a carbon absorber are all used to filter the contaminates. This unit has an output of 58 cfm, and has a noise level of 60DB. The power required are 2.5 amps at 110 volts and 4.4 amps at 12 Vdc. The power requirement is not a problem if the power grid is working. However, a 12 volt battery will be depleted in less than a day if you need to run on battery power.

Can an improvised ventilation system be constructed?

You can improvise your own limited filtration system using residential HEPA filters and a 12 volt fan. The level of protection against radioactive particles will be unknown because these filters have not been designed or tested for this application.

The following is a description of a homemade improvised system that we built for use in our safe room inside the house. The air flow using one computer fan was about 7 cfm. Two fans doubled this to 14 cfm. This was determined by measuring the length of time it takes to fill a large garbage bag. Each fan draws about 0.35 amps of current at 12 volts. You may need two or more of these units to produce the minimum of 5 cfm per person.

The materials needed are:

- 2-foot length of 1x8 board

- 3-foot length of 1x4 board

- Holmes HAPF30 filter

- 4-inch PVC pipe

- Two 120mm by 120mm computer fans

- Tape (to seal seams)

- Screws for attaching boards together

Cut the 1x8 board into two one-foot lengths. Cut a hole in the center of this board that the 4" PVC pipe will tightly fit into. In the other board, cut a rectangular hole for the filter. You will need to cut two notches on each side. These pieces will form the top and bottom of a box.

Next, cut two one-foot lengths from the 1x4 board. Then cut two short lengths of 1x4 boards to complete the sides of the box.

Glue the PVC pipe to the pre-cut board. Use whatever length of PVC you need for your application. You may only need about six inches to go through a wall. If your intake has to go up to reach the ground level, then add an elbow and whatever additional PVC pipe you need. Mount the fan on the other side of this board. The fan only operates in one direction. Make sure the airflow is into the box when mounting the fan. Mount a second fan on top of the first fan to double the airflow. Make sure both fans are pushing air in the same direction.

Push the filter into the other board. The filter is pushed in from the inside of the box. You can also install a pre-filter using rubber bands. Use tape to seal the filter so that no air leaks around it.

Drill a small hole for the wires to power the fan, leading downward. Next, assemble the box using screws so that it can be taken apart to replace the filter. Use more tape to seal seams so there are no air leaks.

Next, drill a round hole in the wall to mount the unit. Use nails, screws, or glue to secure the unit in place. Connect it to a battery and make sure the air is flowing into the room.

To replace the filter, turn off the unit. Remove the tape and unscrew the front board. Install the filter and replace the board.

Some disaster scenarios may cause excessive dust that will clog HEPA filters. A nuclear explosion may fill the atmosphere with dust. In these situations you may need to replace the HEPA filter with something more porous. A furnace filter could be cut to replace the HEPA filter. Cloth could be used to filter larger dust particles. If any of the dust is radioactive, care should be taken when replacing the filter. The filter itself will become radioactive. Put on a mask and avoid touching the used filter with bare hands. Discard it into a sealed bag.

Fallout Shelter Room Ventilation/Filter

A fallout shelter within your existing home allows less extensive preparation and expense. The minimal survival room should filter all air entering the room and create a positive air flow. The ventilation fan for a fallout shelter is essentially the same as the wall-mounted unit described above. Use a concrete drill to make a hole in a concrete block for the 4"PVC pipe. The small 4" hole will limit how much gamma radiation can leak into the room.

If you are installing the ventilation fan in an airtight room you will need an exhaust vent. Use a one-way vent that only allows air to flow out. In addition, cut a 4 inch round piece of air filter and push it into the exhaust pipe at the room vent. This will filter any air that does leak into the room from the exhaust vent.

Can you use available portable air filters for a small shelter room?

A room air purifier can be inserted into a wall to direct filtered air into your survival room. This is a crude installation, so it would be best to buy the material and install it only if necessary. With the right material and tools, you should be able to install this in a couple of hours.

The basic idea is simple. Cut a hole in the wall leading to your survival room and install a small room air purifier in the wall. Use foam and duct tape to prevent leaks. This is a makeshift simple solution you can do when moving into the fallout shelter temporarily.

We haven't tried it, but the Holmes HAP242 looks like it would work. This is a small desktop unit that should fit into a wall. The 12.5 inch width should fit between 16 inch centered studs. The air outlet is on the top, so this side needs to be inside the secure room. Readers can shop for air filtration appliances and select a unit that you could keep for an emergency.

To mount this unit in a wall, first identify where the studs are and drill a hole for the purifier to fit into. You may need to build a shelf to stabilize the unit. Fill the cracks around the unit with foam and duct tape.

What are other considerations for fallout shelter ventilation?

The fallout shelter ventilation needs to be separate from the house HVAC (heating, ventilation and air conditioning system). It also needs to be able to operate independent of the power grid. A 12 volt deep cycle marine battery should be used to power the ventilation fan so that it will continue to run if the AC power fails. Some provision needs to be made to recharge the battery. A battery maintainer should be connected while it is in storage.

Depending on the level of contamination in the air, you may be able to leave the shelter for a brief time. Use a radiation meter to determine the risk of leaving when fallout is reduced. If you do not have a meter, you will have to depend on government reports and your own judgment. As an added precaution, hold a wet cloth to your nose and mouth.

How often you can leave the shelter room within the house, to other rooms, may depend on the status of your house. If windows are broken, and your house has been exposed to outside air, the shelter room is your only safe refuge. However, if your house is intact, and you have electrical power to run the HVAC system, you may be able to use the house much of the time. The survival room, more air tight and with a positive airflow, should be safer. It may be wise to spend as much time as possible in the safe room to minimize exposure risk. On the other hand, confining a large family to a small room may not be healthy either. Just use your best judgment.

Any time you do leave the shelter room, make sure it is set up for immediate use if necessary. Maintain the positive airflow so that the room doesn't get contaminated.

Chapter Fifteen

Survival Shelter Supplies

Survival shelter preparation must consider how many people will be at the shelter during a radiological emergency. You can prepare a shelter in your home with supplies for those who are home at the time of the radiation emergency. Those household members who are traveling or at work should carry an emergency radiation kit including a detector, neutralizers, decontamination wipes and protective equipment.

What supplies are necessary for the stocked fallout shelter?

The following are essential needs based on a 10 day shelter period for two people.

Electronics:

Crank powered emergency radio, CB radio

Ventilation system, commercial or improvised

HEPA Air Purifier

Radiation Detector, commercial or supplies for the Kearny Fallout Meter, home-made detector

Hardware:

Fire extinguishers

Duct tape

2-Plastic 5-gallon buckets with lids. – Two for fresh water, (at least one gallon/per person/per day.)

Plastic 5 gallon bucket with lid- for a makeshift toilet with garbage bag liners. (seats are available at camping supply stores)

Cat box litter

Contractor bags or strong garbage bag liners for improvised protective garment and waste, sealed and removed from shelter daily.

Wet wipes- 500 for each person. These will be used for bathing, decontaminating, hand cleaning.

Caulk for sealing cracks.

Furnace filters that fit windows.

Disposable plastic gloves

Bleach to replenish purification of fresh water. (8 drops/gallon water)

Chlorine test strips 0-100 ppm. Drinking water should be at 2-4 ppm.

Respiratory Mask

Candles

Flashlight

Plastic drop cloths

Hammocks

Blankets

Pepper spray

Local maps

Tools, crowbar, shovels, hammers, screwdrivers, saws

Pharmacy:

Wet wipes

Dental hygiene- toothbrush, toothpaste, floss

Personal hygiene- hair brush, baby shampoo, deodorant, razors, toilet paper

Bandages

Medications;

Prescription medicines

Triple antibiotic ointment

Providone-iodine 10% solution as topical disinfectant

Electrolyte re-hydration fluids

Aspirin or NSAIDs

Antihistamine (allergy pills)

Multivitamins, anti-oxidant supplements

Radiation neutralizing agents approved by your physician— for use when advised by authorities or your physician

Gastrointestinal treatments – anti-diarrhea, anti-vomiting, antacids

Respiratory treatments- expectorants, decongestants

Disposable flatware and dishes

Paper, pencils

Grocery:

Non-perishable food –Canned two pounds per person per day

Can opener

Bottled water

Green or white tea bags

Matches

Books:

Medical and first aid handbooks

Holy Bible

Those who travel to work near a nuclear power plant or near a potential nuclear strike target should carry a radiation emergency kit for short term needs.

The emergency kit should include;

Radiation monitor

Baby wipes for decontamination

Potassium iodide and other neutralizers available without a prescription and approved by your physician.

Bottled water

Holy Bible

Packaged food

Crank powered radio

Respiratory masks

Duct tape

Heavy garbage bags

Tools

First Aid supplies

If you can leave a stocked kit in your car or office, you can increase the weight and size and expand the number of items from the shelter list. Non perishable food, radiation neutralizers, bottled water and first aid supplies can be increased.

Chapter Sixteen

EMP and Radiation – the Double Disaster

What is an EMP?

An EMP (electromagnetic pulse) can be created by a nuclear detonation. It can destroy electronics for miles and damage electrical power transformers and circuits miles away from the blast.

How is response to an EMP different than response to a nuclear emergency?

An EMP does not cause immediate life threatening damage. There is time to retreat to your pre-stocked shelter. The concern is the hungry and thirsty refugees who are otherwise healthy. You need to have a well stocked shelter. There is a potential for mob violence. You need to be able to defend your shelter.

However, in a radiation emergency, you have to find shelter immediately "shelter in place." There may be no supplies of food and water available where you are sheltered. It is dangerous to seek food or water outdoors because of fallout.

In an EMP emergency, the outdoor environment is not toxic, but you will have to remain sheltered until the risk of mob violence ends, possibly lasting months.

However, in a radiation emergency, toxic outdoor fallout will dissipate in days to weeks. Mob violence is less likely because of physical injuries and acute radiation sickness.

In an EMP event, electronics are damaged and the power grid, transportation and communications will probably be out of service for months. Survivors must rely on primitive lighting, heating, cooking and communication, unless they have pre-planned. Emergency assistance may not be available for weeks.

However, in a radiation emergency, communications, transportation and power may still be available.

In an EMP event, acute injuries are not expected, but long term malnutrition and infectious disease are a risk based on lack of sanitation systems.

However, a radiation emergency can cause acute physical injuries and deaths. Mortality from acute radiation sickness will occur in the first few days or weeks.

In a nuclear attack both events could occur!

Preparation will require two separate survival strategies. We recommend that you also read <u>EMP Survival: How to Prepare Now and Survive When an Electromagnetic Pulse Destroys Our Power Grid.</u>

Is there a way to prevent the damage to electronics?

A Faraday cage will prevent the electromagnetic pulse from destroying your electronic equipment. Faraday cage construction is described in the book titled above. You can build a simple Faraday cage using two cardboard boxes inside each other, and the electronic device inside the smallest. The outer box should be covered with overlapping aluminum foil, and taped securely. A metal box insulated with foam inside will also function as a Faraday cage.

Chapter Seventeen

Natural Radon

What can you do about natural radiation in your home?

Radon is a natural, but dangerous ionizing radiation source. Unlike nuclear detonation or power plant accidents, humans did not create this risk. The dangerous radiation seeps into homes from underground and can be inhaled or ingested by the occupants. It is the second-leading cause of lung cancer in the U.S.

Radon is estimated to cause up to 21,000 deaths each year, in the U.S. primarily from lung cancer[149]. Radon comes from the natural (radioactive) breakdown of uranium in soil, rock and water. Radon can be found all over the U.S. in homes, offices, and schools. Nearly 1 out of every 15 homes in the U.S. is estimated to have elevated radon levels.[150]

---

[149] Committee on Health Risks of Exposure to Radon, BEIR VI, National Research Council, Commission on Life Sciences, Board on Radiation Effects Research, National Academies Press, 1999

[150] US EPA, "A Citizen's Guide to Radon, Jan 2009, http://www.epa.gov/radon/pubs/citguide.html

Fortunately, radon levels can be effectively reduced with venting. The radon level can be reduced by up to 99% by adding a ventilation system designed for radon. [151]

Radon enters a house through cracks in the walls and floors, gaps around service pipes, or the water supply. Better construction techniques can reduce, but not eliminate the radon.

Well water is more likely contaminated than surface water sources. A deep well originates where radon may be present. The radiation in the water can be aerosolized in the shower and faucet heads and inhaled. 1-2 % of the radon in the air comes from drinking water. [152] It takes 10,000 pCi/L of radon in the water to raise the average radon levels in the house 1 pCi/L[153].

Radon is nine times heavier than air. This is the reason radon builds up in the basement, or other low points in a residence. If you have a home office or living area in your basement, where you spend a lot of your time, radon should be a particular concern.

---

[151] US EPA, "Consumer's Guide to Radon Reduction", September 2010, www.epa.gov/radon/pubs/consguid.html

[152] US EPA, "Basic Information about Radon in Drinking Water", www.water.epa.gov/lawsregs/rulesregs/sdwa/radon/basicinformation.cfm

[153] Leker, "Radon in Water" Water Quality and Waste Management", North Carolina Cooperative Extension Service, March 1996, www.bae.ncsu.edu/programs/extension/publication/he396.html

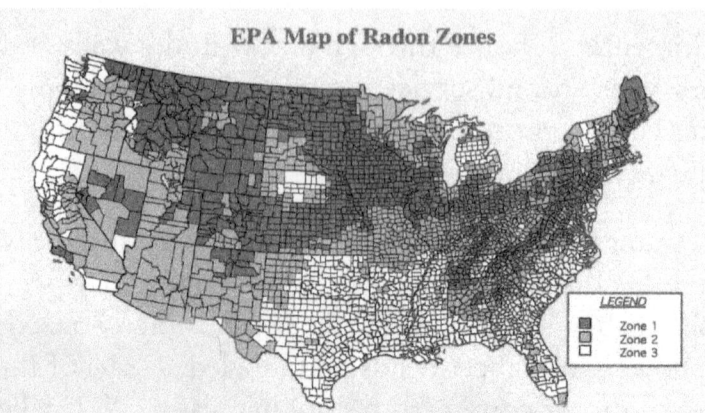

EPA Map of Radon Zones

LEGEND
Zone 1
Zone 2
Zone 3

What are safe levels of radon?

The Congress set a long term goal that indoor radon levels should be no more than outdoor levels. The national average outside radon levels is 0.4 pCi/L. This goal is not yet technologically achievable in all cases, although most homes can be reduced to 2 pCi/L.or below. [154]

The EPA recommendation for a safe radon level is 4 pCi/L. (ten times higher than outdoors) in the home.

The average indoor radon concentration is about 1.3pCi/l of air. It can be found in a range of 5-50 pCi/l. Even 2000 pCi/l has been reported. [155]

How do you test for radon in the air?

---

[154] US EPA, "A Citizen's Guide to Radon", January 2009, www.epa.gov/radon/pubs/citguide.html

[155] US EPA, "Radon", Radiation Protection www.epa.gov/radiation/radionuclides/radon.html

Geiger counters or radiation detectors do not work for radon. A long-term radon test kit is the most reliable. You can evaluate the contamination rate based on a range of seasonal temperature and humidity changes. You can buy a home test kit that you send to a private laboratory for about $20.

A better choice, particularly if you are in a high-risk area, is to buy a radon meter. This allows you to test the radon level in pCi/L anytime you want. Most units include an alarm to warn you of high radon reading. The EPA recommends making venting improvements for radon levels above 4 pCi/L[156]. We believe you should consider a ventilation improvement at a much lower level.

A short term radon test can give you an idea of the radon risk when buying a new house. Many times a real estate transaction will require a radon test. In addition, you should do your own radon test after moving in.

How do you test the water for radon?

Radon test kits for water can be purchased and sent to a private laboratory for testing. You should test water for radon if you have a private well or if you measured radon in the air. Check with your municipal water provider to determine the levels in the public water supply. This could be a concern if the municipal water source is from deep wells.

How do you remedy high radon levels indoors?

It is best to have a qualified radon mitigation contractor make venting changes to your home because they have

---

[156] US EPA, "A Citizen's Guide to Radon", Jan 2009
http://www.epa.gov/radon/pubs/citguide.html

specific technical knowledge and special skills. Without the proper equipment or technical knowledge, you could actually increase your radon level or create other potential hazards and additional costs. Also, by contracting, you will have documentation that the radon problem has been corrected when you sell your house.

For homes with a basement, an active suction is usually the most reliable radon reduction method[157]. One or more suction pipes are inserted through the floor slab into the crushed rock or soil underneath. The radon gas is vented to the outside before it can build-up in the basement. An active system, equipped with fans, is most effective. A passive system with vents only, may be adequate for some homes. Sometimes the radon can be drained through the hole provided for the sump-pump.

For homes with a crawl space, an effective method is covering the earth floor with a high-density plastic sheet. A vent pipe and fan can be used to draw the radon from under the sheet and vent it to the outdoors. In addition to venting radon gas, the level can be reduced by filling cracks and other openings in the foundation.

Completely waterproofing your basement may help to reduce or eliminate radon. One company clams that their waterproofing will completely eliminate radon – not just reduce the level. Other companies offer do-it-yourself waterproofing and radon elimination products. The test for any of these solutions is to measure the radon level yourself to verify the reduction or elimination of a radon problem.

---

[157]US EPA, "A Citizen's Guide to Radon", Jan 2009
http://www.epa.gov/radon/pubs/citguide.html

You can also filter radon from the air using an activated carbon filter. A large high-quality air purifier designed to remove gaseous contaminates is needed. These are costly and noise level should be considered when placing the units. An air filter should be your second level of protection against radon. It is much more effective to eliminate radon where it enters the home.

How do you reduce radon levels in water?

There are three methods to remove radon from water. Eliminating radon from a well can be done with a specialized aeration system. An aeration system is typically installed outside in a vented chamber. Air is injected into the water. The air and the radon are then vented to an exhaust vent on the roof.

A second option to reduce radon from water is to use a granular activated carbon (GAC) system[158]. GAC systems are usually less expensive than aeration systems, but can require special disposal for the filters. Radon accumulates in the filter and can cause exposure to radiation.

A third option is a point-of-use system. These units are installed at each water tap. A point-of-use filter could be used in addition to an aerator or GAC system, but should not be your only protection. Your dishwasher, clothes washer, ice maker, and other appliances will not be protected.

---

[158] US. EPA, "Basic Information about Radon in Drinking Water", February 2011, www.waterepa.gov/lawsregs/rulesregs/sdwa/radon/basicinformation.cfm

Appendix A: The BEIR Reports

The BEIR reports on radon and low level radiation effects were described previously. The reports are considered the authoritative source for radiation and health information. BEIR stands for the "Biological Effects of Ionizing Radiation". The BEIR committees produced the first of seven reports in 1972. They were delegated by Congress through the National Research Council, Commission on Life Sciences. The reports are published by the National Academy Press.

The first BEIR report of 1972 addressed, "The Effects on Populations of Exposure to Low Levels of Ionizing Radiation".

The BEIR II report of the same title was published in 1979 and was considered a draft report.

BEIR III was a final report of the same title, published in 1980.

In 1988, the BEIR IV report, "Health Risks of Radon and Other Internally Deposited Alpha-Emitters", was published.

In 1990, the BEIR V report "Health Effects of Exposures to Low Levels of Ionizing Radiation", was published.

In 1996, BEIR VI, "Health Effects of Exposure to Radon", was published.

In 2006, BEIR VII "Health Risks from Exposure to Low Levels of Ionizing Radiation" was published.

Appendix B: Radiation Information and Survival WEB sites

www.nap.edu/openbook.php?record_id=5499(BEIR VI)

www.afrri.usuhs.mil/www/index.html

www.americansaferoom.com

www.antirad.com

www.atsdr.cdc.gov

www.cdc.gov

www.cddc.vt.edu/host/atomic/pdf/kfm_inst.pdf

www.cedr.lbl.gov

www.crcpd.org

www.dhs.gov

www.emergency.cdc.gov/radiation

www.energy.gov (US Dept of Energy)

www.fbi.gov

www.epa.gov

www.fairewinds.com

www.fema.gov/home/radiolo.htm

www.flu.gov

www.hps.org

www.hps.org/hsc/documents

www.ki4u.com (Kearny Fallout Meter plans)

www.KitsForSurvival.com

www.labsafety.com

www.nationalterroralert.com

www.nilesema.com

www.nrc.gov

www.nrt.org (National Response Team)

www.oism.org (Oregon Institute of Science and Medicine)

www.orise.orau.gov/reacts

www.orau.gov/reacts (Radiation Emergency Assistance Center Training Site)

www.pandemicflu.gov

www.radiation.org

www.Redcross.org

radefx.bcm.tmc.edu

www.radres.org/intro.htm

www.rerf.or.jp

www.remm.nlm.gov/nuclearexplosion.htm

www.rrbike.freeservers.com

www.secretsofsurvival.com

www.survivalunlimited.com/deepwellpump.htm

www.tacda.org

www.urbansurvivaltools.com

www.who.int/ionizing_radiation/en (World Health Organization)

emergency.cdc.gov/radiation/ki.asp

www.bt.cdc.gov/radiation

www.emergency.cdc.gov/radiation/emergencyfaq.asp

www.unscear.org/docs/reports/2008/11-80076-Report_2008_

www.phe.gov/Preparedness/planning/playbooks/stateandlocal/nuclear/pages/background.aspx

Here is a list of survivalist Blogs:

- BackwoodsSurvivalBlog.com

- ForagingPrepper.Blogspot.com

- OffGridSurvival.com

- SuburbanPrepper.com

- SurvivalBlog.com

- SurvivalPreparednessBlog.com

- TheSurvivalistBlog.net

- TheUrbanSurvivalist.com

Appendix C: Books Recommended

There are several books on preparing for a nuclear war. Many of these are out of date, but still contain useful information.

Nuclear War Survival Skills by Cresson H. Kearny published in 1986 is a good reference. This book includes practical information on building fallout shelters. Design plans for the Kearny Fallout Meter and the Kearny Air Pump.

The U.S. Armed Forces Nuclear, Biological and Chemical Survival Manual published in 2003 by Dick Couch and John Boswell, provides an updated description of survival skills. This book discusses the more likely scenarios in today's world: a "dirty bomb", limited nuclear war by new nuclear powers, or a terrorist attack.

The Aftermath, The Human and Ecological Consequences of Nuclear War, published in 1983, by Jeannie Peterson, models the flow of radiation fallout after a nuclear war.

EMP Survival: How to Prepare Now and Survive When an Electromagnetic Pulse Destroys Our Power Grid, by Larry and Cheryl Poole is the first book in our survival series. EMP survival is completely different than radiation survival. The population will not have physical injuries but will lack necessities of modern living. There will be no food or water or power. Mob violence is expected. This is a practical book with specific ideas you can use to prepare for an EMP. It is very possible a nuclear attack would cause an EMP.

The Survival Guide – What to do in a Biological, Chemical or Nuclear Emergency, by Angelo Acquista, gives a detailed description of biological, chemical and nuclear agents

When All Hell Breaks Loose: Stuff You Need to Survive When Disaster Strikes, by Cody Lundin gives practical information on survival along with some humor. This 450 page book is a wealth of information.

Complete Survival Manual: Expert Tips From Four World-renowned Organizations Plus Survival Stories of National Geographic Explorers, by Michael S. Sweeney covers many environments including, forests, mountains, and the desert. You never know where you will be when an attack occurs. Reading through this book will help you prepare.

The facts for this text were derived from reports by many government agencies at the State, Federal and International levels. Footnote references provided in the text will lead the reader to additional information on the topics.

The primary source on fallout survival was the "Planning Guidance for Response to a Nuclear Detonation". This document is available at a number of government websites including http://hps.org/hsc/documents/planning and www.remm.nlm.gov.

The modeling experiments for shelter in place and evacuation were conducted by scientists at Lawrence Livermore National Laboratories. The "Planning Factors for the Aftermath of Nuclear Terrorism" document can be found at www.remm.nlm.gov/ResponsePlanning-LLNL_TR_410067.pdf.

The source of the traditional medical treatment of radiation emergencies was "Medical Aspects of Radiation Incidents". The document can be found at the REAC/TS Radiation Emergency Assistance Center, www.orise.orau.ov/reacts.

The history of radiation accidents was obtained from the Agency for Toxic Substances and Disease Registry of the CDC. The documents can be found at www.atsdr.cdc.gov.

Information on radiation accidents is also available at the U.S. Nuclear Regulatory Commission at www.nrc.gov.

Chernobyl radiation accident information was found at the UNSCEAR, the UN Scientific Committee on Effects of Atomic Radiation, and IAEA, International Atomic Energy Agency. www.iaea.org

Appendix D: Recommended Emergency Radiation Kit

Radiation Detector- commercial unit or materials needed for construction of the Kearny Fallout Meter

Bottled Water

Duct tape

Heavy Plastic garbage bags

Potassium iodide

Respiratory Mask

Wet wipe cloths for decontamination

Neutralizing Agents (if approved by your physician)

Prepackaged food

First aid supplies and books

Holy Bible

Crank powered emergency radio

Tools

See the supplies for a stocked shelter for additional shelter items.

If you can store an emergency kit at your office and in your car you can add to the quantity of supplies.

Appendix E: Radiation Symbols

You may need to interpret the symbols reported by the authorities to know which element you have been exposed to. The following are some of the acronyms of the elements.

Americium	Am	Palladium	Pd
Argon	Ar	Plutonium	Pu
Barium	Ba	Polonium	Po
Beryllium	Be	Radium	Ra
Bromine	Br	Radon	Rn
Cadmium	Cd	Strontium	Sr
Californium	Cf	Technetium	Tc
Carbon	C	Tellurium	Te
Cesium	Cs	Thallium	Tl
Cobalt	Co	Thorium	Th
Gallium	Ga	Uranium	U
Iodine	I	Vanadium	V
Iridium	Ir	Xenon	Xe
Neptunium	Np		

Appendix F: Radiation Measurement Conversions

The dose equivalent measurement for Sv and the rem are computed as follows:

Dose Equivalent = Absorbed Dose x Quality Factor

Quality factor adjusts the value to account for more destructive particles. The quality factor for some common radiation agents follow:

- X-ray or gamma rays   1
- Beta particles   1
- Neutrons (unknown energy)   10
- Protons (unknown energy)   10
- Alpha particles   20
- Particles of unknown energy   20

If you work in a nuclear power plant or medical facility, you may be exposed to particles with a high-quality factor. Plutonium 238 emits alpha particles and produces a high-quality factor in the formula.

For most other situations, you can assume a quality factor of approximately 1. With this assumption, the following can be calculated:

1 R = 1 rem = 1 rad = 0.01 Sv = 0.01 Gy

1 Gy = 1 Sv = 100 rem = 100 R = 100 rad

Convert between units using the following table:

- 1 rem = 0.01 Sv = 10 mSv

- 1 mrem = 0.01 mSv = 10 μSv

- 1 Sv = 100 rem

- 1 mSv = 100 mrem = 0.1 rem

- 1 μSv = 0.1 mrem

www.ingramcontent.com/pod-product-compliance
Lightning Source LLC
Chambersburg PA
CBHW071423170526
45165CB00001B/372